# The Organic Chemistry of Isotopic Labelling

# The Organic Chemistry of Isotopic Labelling

**James R. Hanson**
*Department of Chemistry, University of Sussex, Brighton, UK*

RSCPublishing

ISBN: 978-1-84973-188-1

A catalogue record for this book is available from the British Library

Published by The Royal Society of Chemistry,
Thomas Graham House, Science Park, Milton Road,
Cambridge CB4 0WF, UK

Registered Charity Number 207890

For further information see our web site at www.rsc.org

Printed and bound in Great Britain by CPI Antony Rowe, Chippenham and Eastbourne

# Preface

Isotopically labelled organic compounds play a central role in many chemical, spectroscopic and biochemical investigations, in studies of drug metabolism and in medical imaging. The chemical synthesis of these compounds is an essential prerequisite to the success of these studies. It represents a major point of contact between the organic chemist and the biochemist or physical chemist. The constraints on the synthetic strategy that are posed by the need to label efficiently and economically a specific site in a molecule, often within a particular time-scale and with a defined scientific objective, determine the synthetic route that can be used. In order to fulfil these objectives, the synthetic approach to a labelled compound often differs substantially from the routes that have been used for the conventional syntheses of unlabelled material. Although labelled compounds have provided a valuable insight into mechanistic organic chemistry, it is also true to say that a thorough understanding of organic reaction mechanisms underpins the design of labelling strategies.

Many novel methods have been developed for specifically labelling compounds in order to obtain particular chemical or biochemical pieces of information. Whereas the synthesis of the unlabelled material may be predicated by dissections that are based on the relationships between functional groups, the synthetic route to the labelled material may be determined more by the availability of specifically labelled starting materials and the chemistry that is required to introduce them. In order for there to be a fruitful collaboration between the biochemist and chemist, each needs to be aware of the range of methods that are available for labelling compounds and the constraints that are involved.

The Organic Chemistry of Isotopic Labelling
By James R. Hanson
© James R. Hanson 2011
Published by the Royal Society of Chemistry, www.rsc.org

It is the object of this book to describe the various synthetic routes for labelling compounds.

There has been an interaction between the development of novel isotopic methods and the solution of many chemical and biochemical problems. In evaluating progress in these areas, it is important to appreciate the methods that were available at the time at which the work was carried out and thus to consider the experiments in their historical context. Hence the first chapter contains a brief historical introduction to the chemistry of isotopic labelling before describing some general criteria which affect the choice and detection of a label. Subsequent chapters deal with the specialized chemistry of carbon and hydrogen labelling. The concepts of the conformational analysis of reactions which were developed in the 1950s and 1960s had a major influence on the prediction of the stereochemistry of labelling compounds with deuterium and tritium and in the application of stereospecifically labelled compounds to chemical and biochemical problems. Aspects of these topics, including the labelling of amino acids, are discussed in Chapters 4 and 5. The development of a new medicine requires a thorough understanding of the way in which it works, its distribution and its metabolism. Together with legislation on the safety of medicines and the forensic analysis of drugs of abuse, this has generated a considerable requirement for labelled compounds. Some examples of the labelling strategies used in preparing these compounds are described in Chapter 6. The introduction of other stable and radioactive isotopic labels forms the subject of the subsequent chapters. The final chapter describes the synthesis of isotopically labelled compounds for use in medical imaging. Because of the short half-life of the isotopes concerned, this area has provided a considerable challenge to the synthetic chemist.

Only the salient steps are shown in the reaction schemes to provide a guide to the overall strategy. Where appropriate, sites that are labelled are indicated by an asterisk. However when a compound has been labelled at several sites by different routes and the same structure is being used, in order to avoid confusion, this has not been done. In the majority of these cases the relevant sites are numbered.

The book is derived from 50 years of experience working with labelled compounds. It is aimed at the chemist or biochemist who is about to embark on a research project involving the preparation of isotopically labelled compounds. It will also be useful to those who have to teach organic chemistry, particularly to biochemists, by providing fairly simple examples of modern organic chemistry and the use of reaction mechanisms in designing syntheses.

I wish to thank Professor Christine Willis of the University of Bristol and Dr Merlin Fox of the Royal Society of Chemistry for their helpful comments on the manuscript.

J. R. Hanson

# Contents

The Organic Chemistry of Isotopic Labelling
By James R. Hanson
© James R. Hanson 2011
Published by the Royal Society of Chemistry, www.rsc.org

# An Introduction to the Organic Chemistry of Isotopic Labelling

**Chapter 1** The discovery and detection of isotopes.

The aim of this chapter is present a brief history of the impact of isotopic labelling on the development of organic chemistry and biochemistry. Some general principles are described concerning the selection of an isotopic label for specific purposes and the underlying physical methods for their detection. A summary of the nomenclature of isotopically labelled compounds is given.

**Chapter 2** Labelling compounds with carbon-13 and carbon-14.

The aim of this chapter is to describe and exemplify the major methods for introducing a carbon-13 or carbon-14 label into an organic compound. These include methods based on the use of the cyanide ion, organometallic methods including the Grignard and Wittig reactions and various carbonyl condensation reactions. The synthetic routes are exemplified by the synthesis of biosynthetic intermediates such as mevalonic acid and various aromatic and heterocyclic systems such as the nucleic acid bases and intermediates in alkaloid biosynthesis.

**Chapter 3** Labelling with deuterium and tritium.

The aim of this chapter is to describe general and site-specific methods for introducing deuterium and tritium into organic compounds. The role of the kinetic isotope effect is described. Methods which involve exchange reactions on aromatic rings and adjacent to carbonyl groups and also decarboxylation reactions are exemplified. Reduction,

The Organic Chemistry of Isotopic Labelling
By James R. Hanson
© James R. Hanson 2011
Published by the Royal Society of Chemistry, www.rsc.org

hydrogenolysis, hydroboration and organometallic methods are summarized.

**Chapter 4** Stereochemical aspects of labelling with hydrogen isotopes.
The Cahn–Ingold–Prelog sequence rules for designating the stereo-chemistry of chirally labelled centres are outlined. The stereochemical outcomes of reactions used to introduce a label particularly into cyclic systems are outlined. These are exemplified by the stereospecific introduction of labels into the steroids and gibberellin plant hormones. Methods for the preparation of the chiral methyl group and various chirally labelled biosynthetic precursors such as mevalonic acid, isopentenyl diphosphate and 1-deoxy-D-xylulose are discussed, in addition to the determination of the stereochemistry of hydrogen transfer from the nicotinamide coenzymes.

**Chapter 5** The synthesis of labelled amino acids.
The syntheses of amino acids containing both stable isotopes and radioisotopes are described. The methods are divided into those which lead to the racemic amino acid and those which lead to enantio-merically pure amino acids. A number of the methods involve a combination of chemical and enzymatic methods.

**Chapter 6** The labelling of some compounds of pharmaceutical interest.
The isotopic labelling of pharmaceutical compounds is an important requirement for establishing their metabolism. Strategies for intro-ducing labels are exemplified by lidocaine, some non-steroidal anti-inflammatory agents, amphetamines, sulfonamides, morphine and its relatives, galanthamine, tropane alkaloids and tryptamine derivatives.

**Chapter 7** Labelling compounds with the stable isotopes of nitrogen and oxygen.
The organic chemistry of labelling of amino acids, nucleic acid bases and other heterocyclic compounds with the stable isotope nitrogen-15 is described. The introduction of oxygen-17 and oxygen-18 into com-pounds by nucleophilic substitution and the hydrolysis of acetals is re-viewed. The preparation of chiral $[^{16}O,^{17}O,^{18}O]$phosphate is described.

**Chapter 8** Labelling compounds with isotopes of phosphorus, sulfur and the halogens.
The organic chemistry of labelling compounds with radioactive isotopes of phosphorus, sulfur and chlorine is reviewed. The use of radio-isotopes of iodine as therapeutic radiopharmaceuticals and in radio-immunoassay is reviewed.

**Chapter 9** Labelling organic compounds for diagnostic imaging.

The organic chemistry of labelling compounds for use in diagnostic imaging techniques such as positron emission tomography (PET) and single photon emission computed tomography (SPECT) is reviewed. Methods for the rapid introduction of carbon-11, fluorine-18, nitrogen-13, oxygen-15, technetium-99m and iodine-123 for these purposes are described.

The book ends with some conclusions and recommendations for further reading concerning the material in the individual chapters together with a glossary of terms used in isotope chemistry.

CHAPTER 1

# The Discovery and Detection of Isotopes

## 1.1 INTRODUCTION

Labelled compounds in which a specific atom has been replaced by an isotope have played a major role in elucidating reaction mechanisms in chemistry, in establishing metabolic pathways in biochemistry and in imaging organs in medicine. This book is concerned with the organic chemistry that has been used in the synthesis of isotopically labelled compounds employed in these studies.

The isotopes of an element have a nucleus with the same number of protons but vary in the number of neutrons that are present. Each isotope of an element has the same atomic number but a different atomic mass. Since the different isotopes of the same element have the same number of protons and electrons, their overall chemical properties are virtually identical, although there may be subtle differences associated with reaction rates. However, the physical properties of isotopes can differ sufficiently to allow their presence to be detected. Some isotopes such as deuterium and carbon-13 are stable whereas others such as tritium and carbon-14 are unstable and undergo radioactive decay. Since isotopes can be detected by differences in their physical properties, they can act as tracers for the origin and fate of particular atoms within a molecule in the course of a reaction. The application of isotopic tracers in chemistry, biochemistry and medicine has led to the solution of many problems of wide interest.

The Organic Chemistry of Isotopic Labelling
By James R. Hanson
© James R. Hanson 2011
Published by the Royal Society of Chemistry, www.rsc.org

## 1.2   THE DISCOVERY OF ISOTOPES

The discovery, isolation and utilization of isotopes took place throughout the twentieth century. As their availability and the methods for their detection became more refined, so the nature of the problems that could be examined by isotopic methods changed. After the discovery of X-rays by Röntgen in 1895, the first physical observations of radioactivity were reported in 1896 by Becquerel. In 1898, studies by Marie Curie and her husband Pierre Curie on the uranium ore pitchblende led to the isolation of the radioactive elements polonium, a congener of bismuth, and radium, which was obtained from the barium fraction. Further radioactive elements were then discovered associated with thorium.

Investigations by Rutherford in 1899 into the nature of the radiation emanating from these materials and its response to magnetic fields revealed the presence of α- and β-rays. The more penetrating γ-rays ware detected by the Curies in 1900 in the radiation from radium atoms. The distinction between α-, β- and γ-rays was not only in their deflection by electromagnetic fields but also in terms of their penetrating power. At this time radiation was detected not just by photographic plates but also by the flashes produced by a phosphor, zinc sulfide, a precursor of modern scintillation counting. Early versions of the Geiger counter, which relied on the ionization of a gas permitting an electric discharge, were introduced in 1908. The Wilson cloud chamber was introduced in 1911 and this revealed the track of an α-particle and β-radiation by the condensation of supersaturated moist air on an ionized gas. Many of the physical foundations for the detection of radioactivity were laid before the First World War.

The ideas of radioactive disintegration and radioactive decay followed from the work of Rutherford, Soddy and Becquerel in the early years of the twentieth century. The identification of disintegration or decay series with a non-radioactive end product then followed. The end product is typically lead. A consequence of lead being produced from different decay series was that the measured atomic weight and isotopic distribution of lead vary from one source to another. This has been used in the modern environmental monitoring of the source of lead contamination.

The construction of these decay series led to the term *isotope* being coined by Margaret Todd and Frederick Soddy in 1913 to denote atoms of the same element having different nuclear masses. Many of the radioactive isotopes that were detected at this time were of the heavier elements.

A precursor of the mass spectrometer, which revealed the presence of isotopes of lighter elements, was described by Thomson in 1912. A gas was ionized in a cathode ray tube and passed through electric and magnetic fields. The ions were deflected and collected on a photographic plate. When neon was examined in this tube, two spots appeared, which were identified as neon-20 and neon-22. In 1913, and then after the First World War, Aston was able to show that these isotopes could be fractionated by careful distillation and by diffusion through a clay pipe, thus paving the way for later diffusion methods of separating isotopes.

The development of the mass spectrometer by Aston from 1919 through the 1920s, and with various modifications by Dempster and Bainbridge in terms of focusing methods and by Nier in 1940 on the accelerator beam, provided a valuable instrument for the detection of isotopes. The number of known isotopes increased rapidly through the 1920s. Other investigations during the 1920s by Mulliken provided evidence for the existence of a number of isotopes of the light elements leading to the detection of species such as $^{16}O-^{18}O$, $^{15}N-^{16}O$ and $^{13}C-^{16}O$.

Developments during the 1930s in the preparation of isotopes laid the foundations for their application as tracers. Methods were found for the enrichment of stable isotopes and the creation of radioactive isotopes of the lighter elements that were of interest to the organic chemist and biochemist. Many of the first examples are found in the 1930s in which isotopic tracers were used in the elucidation of organic reaction mechanisms and in the delineation of biochemical pathways.

Although the existence of deuterium was predicted by Rutherford in 1920, the isotope was not detected by Urey until 1932. In 1923 a suggestion had been made by Kendall and Crittenden that isotopes might be separated by electrolysis, and in 1933 Lewis found that the electrolysis of water led to the concentration of deuterium oxide. A number of other methods of separation were explored at this time, including the distillation of water, the adsorption of hydrogen or water on charcoal, the diffusion of hydrogen through palladium and the reaction of water or acid with metals. However, over the next few years, systematic studies on the electrolysis of water led to the commercial availability of deuterium oxide (heavy water) and to many applications of this isotope.

Exploitation of equilibrium isotope effects in exchange reactions by Urey during the 1930s permitted the enrichment of a number of isotopes in useful quantities. This was applied to the enrichment of nitrogen-15 using the exchange reaction between ammonia gas and ammonium nitrate solution:

$$^{15}NH_3(g) +^{14}NH_4^+(soln) \leftrightarrow ^{14}NH_3(g) +^{15}NH_4^+(soln)$$

The ammonia gas was passed up a column down which a solution of ammonium nitrate was percolating. The nitrogen-15 was concentrated in the ammonium nitrate. Modifications to the construction of the column eventually gave a system which allowed an enrichment to 60% nitrogen-15. The system was modified in 1940 to enrich carbon-13 by using the exchange reaction between gaseous hydrogen cyanide and a solution of sodium cyanide; the reaction was carried out in the presence of sodium sulfite to prevent the polymerization of hydrogen cyanide:

$$H^{12}CN(g) + {}^{13}CN^-(soln) \leftrightarrow H^{13}CN(g) + {}^{12}CN^-(soln)$$

In this case the carbon-13 was concentrated in the gas. It was possible to obtain an enrichment of 25%. Much higher enrichments have been obtained later by other methods.

Water sufficiently enriched in oxygen-18 for chemical studies was produced by Ingold and Watson in 1937 by careful fractional distillation. More recent methods for obtaining these isotopes are described later in the book when the individual isotopes are discussed.

The foundations for the artificial disintegration of elements and the eventual creation of artificial radioactivity were laid in 1919. Rutherford bombarded nitrogen with α-particles and obtained fast-moving protons by the reaction

$$^{14}N + {}^4He \rightarrow {}^{17}O + {}^1H$$

These were detected by a zinc sulfide phosphor with different sheets of mica being placed between the source and the detector. A number of developments were reported in 1932 which paved the way for the discovery of artificial radioactivity. Up to that time, nuclear disintegration had been induced by naturally produced α-particles. Cockroft and Walton reasoned that the smaller hydrogen nuclei had a better chance of penetrating the atom than the α-particle. They ionized hydrogen by electron impact and accelerated the ions across a potential drop. Bombardment of a target such as lithium led to nuclear disintegration with the formation of pairs of helium nuclei:

$$^1H + {}^7Li \rightarrow {}^4He + {}^4He$$

In 1932, Lawrence and Livingston designed the cyclotron as a means of accelerating ions using an alternating current and a magnetic field.

Ions were repetitively accelerated through the alternating electric field to give them sufficient energy to overcome the electrical repulsion of the nucleus.

Two other discoveries in 1932 were of the positive electron known as the positron and the neutron. The latter was discovered by Chadwick as a consequence of its effect on other elements such as lithium. A commonly used source of neutrons became the bombardment of beryllium atoms by $\alpha$-particles produced by the decay of radium:

$$^{9}\text{Be} + {}^{4}\text{He} \rightarrow {}^{12}\text{C} + {}^{1}\text{n}$$

Because of their lack of charge, neutrons have a greater likelihood of penetrating the atom compared with $\alpha$-particles. High-energy neutrons tend to produce nuclear disintegration. However, if the neutron energy is moderated by passage through water or paraffin wax, the neutron may be captured by a nucleus to produce a nuclear reaction.

In 1933, Madame I. Curie-Joliot and her husband F. Joliot were studying the effect of bombarding boron, magnesium and aluminium with $\alpha$-particles. They found that positrons were emitted in addition to protons and neutrons. In 1934, they reported that the positron emission continued after the activating radiation had been removed. The radioactive substances producing this secondary radiation were separated from their parents and shown to be the source of the positrons. Thus irradiation of boron nitride gave a radioactive isotope of nitrogen whereas aluminium gave radioactive phosphorus:

$$^{10}\text{B} + {}^{4}\text{He} \rightarrow {}^{13}\text{N} + {}^{1}\text{n}; \quad {}^{13}\text{N} \rightarrow {}^{13}\text{C} + \text{e}^{+}$$

$$^{27}\text{Al} + {}^{4}\text{He} \rightarrow {}^{30}\text{P} + {}^{1}\text{n}; \quad {}^{30}\text{P} \rightarrow {}^{30}\text{Si} + \text{e}^{+}$$

These reports stimulated many further studies and by the end of 1935 it was estimated that over 100 new radioisotopes had been discovered. Although many of these were short-lived curiosities, there were a number of considerable importance, such as carbon-11. By the mid-1930s, many of the physical phenomena had been reported which were to underpin the development of the organic chemistry of isotopic labelling.

## 1.3 EARLY APPLICATIONS OF ISOTOPICALLY LABELLED COMPOUNDS

The isolation in the 1930s of both stable isotopes and radioisotopes of elements found in organic compounds soon led to their application to biochemical and chemical problems. Comparison of the contents of the *Biochemical Journal* and the *Journal of Biological Chemistry* between the late 1930s and 1940s and a decade earlier reveals the impact of isotopic methods. Prior to this time, many of the metabolic relationships between compounds had been proposed as a result of feeding large amounts of a putative precursor to a living system, with the inherent dangers of perturbing pathways that such a strategy involves. In 1935, Schoenheimer and Rittenberg reported the catalytic deuteriation of linoleic acid to stearic acid and its incorporation by rats into fats. They also prepared a deuteriated sample of the sterol 5β-cholestan-3-one (coprostan-3-one) and used it to show that it was an intermediate in the biological conversion of cholesterol to coprostanol. The presence of deuterium in the metabolites was established by combustion and the use of micro-density methods to determine the density of the water that was produced. In 1937, the same group prepared amino acids labelled with nitrogen-15. The first experiments used [$^{15}$N]glycine and [$^{15}$N]hippuric acid and showed that they were absorbed by rats. Subsequent experiments with [$^{15}$N]amino acids were used in a form of dilution analysis to detect the amounts of various amino acids in protein hydrolysates such as that of α-lactoglobulin. The $^{15}$N:$^{14}$N ratio was measured mass spectrometrically in the ammonia obtained by a micro-Kjeldahl digestion of the amino acid. One of the earliest double-labelling experiments, which was reported in 1940, examined the metabolism of the unnatural amino acid D-leucine to its natural L-enantiomer. The labelled sample contained deuterium in the chain and nitrogen-15 in the amino group. The deuterium was substantially retained in the L-leucine that was produced whereas the nitrogen-15 was lost. The utilization of the methyl group of methionine in the biosynthesis of choline and creatinine was reported by du Vigneaud in 1941 using [$^2$H$_3$-methyl]methionine. The utilization of acetic acid in the biosynthesis of cholesterol was demonstrated by Bloch and Rittenberg in 1942 using deuterium labelling.

Radioactive phosphorus was employed by Hevesey in 1935 to examine the distribution of phosphate in various tissues in the rat and further experiments on phospholipid metabolism using this isotope were reported by Perlman in 1940. The application of radioactive iodine in the examination of thyroid function was reported in 1940.

However, a major impact of radioisotopes on biochemistry came from studies on the fixation of carbon dioxide by photosynthesis and as part of other biosynthetic processes. Initial work, reported in 1939 and 1940 by Ruben and Kamen, utilized [$^{11}$C]carbon dioxide with barley and the green alga *Chlorella pyrenoidosa*, but it was hampered by the short half-life of carbon-11. The introduction of carbon-14 enabled the pathway of carbon in photosynthesis to be established, particularly through the later work of Bassham and Calvin. The products obtained from exposure of the photosynthetic organism to [$^{14}$C]carbon dioxide were separated and identified by paper chromatography and radioautography so that by the 1950s a complete pathway was established. These early applications of isotopes to biochemical problems revealed that there was a dynamic metabolic relationship between various cellular constituents and provided the stimulus for a rapid expansion in the study of the biochemical pathways of metabolism.

Physical organic chemistry also benefited rapidly from the advent of stable and radioisotopes. In 1935, Hughes, in a study of the Walden inversion, established that the rate of exchange of a radioactive isotope of iodine in sodium iodide with the iodine of optically active *sec*-octyl iodide was equal to the rate of racemization of the alkyl iodide, thus linking aliphatic substitution and optical inversion.

Several methods for preparing deuteriobenzenes including exchange reactions with deuteriosulfuric acid and the distillation of the calcium salts of benzenecarboxylic acids with calcium deuteroxide were reported by Ingold and Wilson and others between 1934 and 1936. The products of these reactions were used in assignments of the infrared spectra. In subsequent years, exchange and decarboxylation reactions were used to prepare many more deuterated compounds for spectroscopic assignments. Correlations were published in 1936 between the exchange reactions of phenol and resorcinol and the sites of aromatic substitution.

Another area in which isotopic substitution played an important role in the late 1930s was in work reported by Urey and by Ingold on the fate of oxygen-18 from [$^{18}$O]methanol and [$^{18}$O]water in esterification and ester hydrolysis. The fate of isotopic labels in rearrangement reactions was also used in mechanistic studies of these reactions. Thus, in 1943, nitrogen-15 was used in a study of the mechanism of the Fischer indole synthesis and a classical study of the mechanism of the Claisen rearrangement of phenyl allyl ethers, reported in 1952, examined the fate of carbon-14 in the terminal position of the alkene.

Although the Second World War led to a slowdown in academic research during the early 1940s, the availability of radioisotopes from nuclear research through programmes in the post-War period that were directed at the peaceful uses of radioactivity provided a powerful stimulus to the synthesis of labelled compounds. Thus the Radiochemical Centre (now GE Healthcare) was created after the Second World War in laboratories at Amersham in the UK which had been used in the war effort to produce luminous materials from radium sources. By the end of the 1940s, it was producing carbon-14-labelled compounds. It has been estimated that by the end of the 1960s over 500 labelled compounds were commercially available from these and other laboratories, such as New England Nuclear, founded in 1956 and now part of Perkin-Elmer. During the late 1940s and early 1950s, many papers appeared in the literature describing the preparation of labelled compounds of biochemical interest such as amino acids.

Evidence for the existence of tritium had been presented in 1934 and 1935 and its radioactivity had been described in 1939. After the War it was available from the irradiation of lithium in a nuclear pile in the reaction $^6Li(n, \alpha)^3H$. However because it was a very soft $\beta$-emitter (maximum 0.016 meV), it was relatively difficult to analyse prior to the advent of the liquid scintillation counter. The development of the liquid scintillation counter in the early 1950s, coupled with a greater understanding of the role of isotope effects and the mechanistic information which could be obtained from them, led to an expansion of the role of tritium as an isotopic label. It found particular application in the study of reactions in which a carbon–hydrogen bond is broken, such as in oxidation and in aromatic substitution reactions. The measurement of isotope effects made a major contribution to the identification of the rate-determining steps in these reactions.

During the late 1950s and early 1960s, there was a significant change in the nature of the biochemical problems that were tackled by isotopic labelling experiments. There was a shift from just establishing the pattern in which building blocks were incorporated to answering questions of detailed biochemical mechanisms particularly involving the stereochemistry of enzyme reactions. This required the synthesis of stereospecifically deuteriated and tritiated substrates and is exemplified during the 1960s by the work of Cornforth on the nicotinamide coenzymes, by Cornforth and Arigoni on the mevalonic acids and by Barton and Battersby on alkaloid biosynthesis.

Another significant change in the use of isotopes for labelling studies in biosynthesis came from the development of $^{13}C$ NMR spectroscopy in the 1960s. Studies in the 1950s on the role of acetate units in the biosynthesis of polyketides by Birch and of cholesterol by Cornforth and Popjak required considerable chemical effort in the degradation of the products of biosynthesis in order to demonstrate the specificity of labelling. The assignment of carbon-13 resonances in the $^{13}C$ NMR spectra, particularly of fungal metabolites, and the availability of more sensitive high-field spectrometers brought this spectroscopic method into the realms of biosynthetic studies. Many ingenious experiments were devised by Tanabe, Simpson, Staunton, Arigoni, Cane, Vederas and others during the 1970s and 1980s to exploit this methodology in polyketide and terpenoid biosynthesis. This required the synthesis of substrates specifically labelled with carbon-13 and deuterium. Several monumental biosynthetic studies in the latter part of the twentieth century, such as those by Baldwin on penicillin and by Scott and Battersby on vitamin $B_{12}$, required the synthesis of a wide range of stereospecifically labelled substrates in order to understand not just the sequence of biosynthetic events but also their mechanism. Although $^3H$ NMR proved valuable in the 1970s for demonstrating the specificity of tritium labelling in chemical syntheses, it did not then have sufficient sensitivity to be used routinely for biosynthetic studies where the incorporation of a labelled substrate could be low. Changes in the regulations governing the registration of new medicines, an increasing realization of the consequences of their metabolism and the importance of establishing the pharmacokinetics of new drugs led to considerable efforts in the preparation of labelled compounds. Since the 1960s, isotopic labelling has played a very important part in the pharmaceutical industry in the development of new drugs.

As more proteins have been purified and attention has focused on their folding and the role of interactions between specific amino acids in their constituent chains, so the need has arisen for the synthesis of specifically labelled amino acids, exemplified by the work of Young and Willis. The application of isotopic labelling to the identification of specific NMR signals and their variation in substrate–enzyme interactions can afford valuable mechanistic insight into enzyme catalysis.

The development of non-invasive techniques for medical imaging has had a significant impact on the organic chemistry used in labelling compounds. Positron emission tomography employs a short-lived radioactive tracer which decays with the emission of a positron. The technique is used to map the metabolic activity of cells within the body. These applications, first developed in the 1970s, require the very rapid synthesis of the labelled substrates. This imposes very different constraints on the synthetic organic chemist to those that are met in conventional organic synthesis.

These developments have required many novel syntheses of labelled compounds, which are described in the following chapters.

## 1.4 THE SELECTION OF AN ISOTOPIC LABEL

The common isotopes that have been used to label organic compounds and have found application in chemical and biochemical problems are listed in Tables 1.1 and 1.2. They are divided into radioactive isotopes (Table 1.1) and stable isotopes (Table 1.2). The isotope that is used for

**Table 1.1**   Common radioactive isotopes used in labelling experiments.

| Element | Isotope | Half-life | Emission |
|---------|---------|-----------|----------|
| Hydrogen | Tritium, $^3$H | 12.26 yr | $\beta^-$ |
| Carbon | $^{11}$C | 20.4 min | $\beta^+$ |
|  | $^{14}$C | 5568 yr | $\beta^-$ |
| Nitrogen | $^{13}$N | 9.96 min | $\beta^+$ |
| Oxygen | $^{15}$O | 2 min | $\beta^+$ |
| Fluorine | $^{18}$F | 110 min | $\beta^+$ |
| Phosphorus | $^{32}$P | 14.22 d | $\beta^-$ |
| Sulfur | $^{35}$S | 87.4 d | $\beta^-$ |
| Chlorine | $^{36}$Cl | $3.08 \times 10^6$ yr | $\beta^-$ |
| Bromine | $^{82}$Br | 35.87 h | $\beta^-, \gamma$ |
| Iodine | $^{123}$I | 13.13 h | $\gamma$ |
|  | $^{125}$I | 59.4 d | $\gamma$ |
|  | $^{131}$I | 8.02 d | $\beta^-, \gamma$ |

**Table 1.2**   Common stable isotopes used in labelling experiments.

| Element | Isotope | Natural abundance (%) |
|---------|---------|-----------------------|
| Hydrogen | Deuterium, $^2$H | 0.015 |
| Carbon | $^{13}$C | 1.1 |
| Nitrogen | $^{15}$N | 0.38 |
| Oxygen | $^{17}$O | 0.02 |
|  | $^{18}$O | 0.2 |

labelling a compound is dictated by the application of the labelled compound. The labelled compound may be required in order to establish a sequence of biosynthetic events or to identify the metabolic fate of a drug. It may be employed to reveal the mechanism of a biochemical process and the interactions between a substrate and an enzyme. The purification of an enzyme system can be monitored with a labelled compound. The fate of a labelled compound can establish the stage in the life cycle of a living organism at which a biosynthesis occurs and it can be used to image the site of metabolic activity. Radioactive and stable isotopes have different advantages which may be exploited for these various types of experiment.

Whereas radioisotopes may have the sensitivity that is useful in determining sequences and the time-scale of events, stable isotopes with NMR and mass spectrometric methods of detection may be more useful in establishing the integrity of biosynthetic units. Mass spectrometry coupled with gas chromatography or other separation techniques is a very powerful method of analysis for trace amounts of metabolites. Compounds labelled with stable isotopes can play an important role as internal standards for quantitative work. Monitoring the appearance of a radioactive label such as carbon-14 or tritium in metabolites may provide quantitative information on the sequence of biosynthetic events. A radioactive label is also valuable in trapping experiments. However, a stable isotope such as carbon-13 or deuterium may be more useful for the spectroscopic identification of the eventual site of a label. Prior to the advent of NMR methods, chemical degradation had to be undertaken to locate the site of a label. The ease of this degradation often had a bearing on the site of the label in the substrate.

The aim of the work may involve a study of the kinetics of a chemical or enzymatic reaction, in which case the sensitivity and quantitative nature of a radioactive label such as tritium may be the most useful. Automated instrumentation can be particularly useful in the context of measuring the radioactivity of a number of samples. In such studies, an isotope effect based on comparing the rate of reaction of a tritiated or deuteriated species with that of its protic counterpart may provide additional information on the nature of the rate-determining step in a sequence.

Different organisms may use different pathways to create or metabolize similar compounds. A classic example is that of terpenoid biosynthesis where *Streptomycetes* utilize the deoxyxylulose pathway to create the isoprene unit, whereas fungi and mammals utilize the mevalonate pathway. In plants the deoxyxylulose pathway predominates in the chloroplasts whereas the mevalonate pathway is found in the cytosol. In some instances there is a cross-over between these pathways. This biosynthetic diversity can have consequences when the design of a tracer experiment is being considered.

Access to the biosynthetic enzymes may also determine the way in which the precursors are labelled. For example, there are situations where a substantial portion of the carbon skeleton is created by a multi-enzyme complex, the structure of which precludes the incorporation of some intermediates. This is the situation with polyketide and fatty acid biosynthesis, where the synthases will not accept precursors between acetate and the ultimate long chain. Many ingenious experiments utilizing multiply labelled acetate precursors have been devised to overcome this problem. The use of several isotopes in the same substrate, for example deuterium and carbon-13, may be indicated.

There are some precautions which need to be considered when using an isotopically labelled compound to establish a biosynthetic sequence. First, the amount of labelled material that is used should be a minimum such that it does not perturb the biosynthetic sequence by inducing a step or by creating a metabolic stress which leads to regulatory modification of a pathway. The putative intermediate may be related to a genuine intermediate and readily transformed into it but may not itself lie on the pathway. A metabolic grid of alternative pathways inter-relating various hydroxylated metabolites may exist but with a species-specific dominant route through the grid. Time-course experiments in which the appearance of a label in the putative intermediate from an earlier precursor should be compatible with its incorporation. There is also the possibility that a putative intermediate may be degraded to smaller units before re-synthesis occurs. Retention of the specificity of labelling between substrate and metabolite and the use of multiple labels are helpful in establishing the integrity of a unit in a biosynthetic sequence. The labelled compound may be an adjunct to another aspect of the study such as providing a monitor for the purification of an enzyme system or the binding of a drug to an enzyme or receptor. There may be a need to make measurements on multiple samples. In these cases the sensitivity of a radioisotopic method may be of value for quantitative measurements, particularly if the site of the label within the target molecule is of less importance.

A further group of studies involving labelled compounds are those aimed at imaging of particular tissue within the body. Techniques such as positron emission tomography (PET) and single photon emission computed tomography (SPECT) may be employed. These techniques use isotopes such as carbon-11 and fluorine-18 which emit a positron. On collision with an electron this undergoes annihilation with the emission of two photons. These isotopes have only a short half-life and therefore cause only limited radiation damage to the subject. However, the time that can elapse between their production and final application is very short and imposes a limit on the compounds that can be used and the laboratories that can undertake this work. The much older and cheaper techniques of autoradiography provide useful information on the localization of biosynthesis in plants. Some radioactively labelled substances that target particular organs can be used in radiotherapy.

There are a number of other constraints which determine the isotope that can be used in an experiment. Principal amongst these are cost and safety. There are serious safety implications in the use of radioisotopes. In this context, there is a significant negative public perception of the use of radioisotopes in experimental work. Precautions must be taken which are aimed at protecting the chemist and the environment, minimizing exposure to radioactivity, providing containment of the radioactive hazard. particularly in the event of an accident, and controlling the disposal of radioactive waste after the experiment. The safe storage of radioisotopically labelled compounds, the disposal of radioactive waste and the decontamination of apparatus need to be considered before an experiment is undertaken. There are important precautions that need to be taken which vary with the isotope and the energy of the radiation that it produces. Whereas glass will absorb the radiation from some isotopes, lead shielding is required for others. Special precautions need to be taken with the radioactive isotopes of iodine, which may become localized in the thyroid gland. There are national regulations which have to be followed and most laboratories also have local safety protocols for radioisotopic work. In many laboratories there are compulsory training courses in the safe handling of radioisotopes. Coupled with modern detection methods, stable isotopes have much to commend them.

Having selected the isotope for use in an experiment, there are some general considerations which need to be made when designing the synthesis of a labelled compound. The ease of chemical access to the site of the label is a major criterion. The label should be introduced with the minimum number of steps subsequent to the labelling reaction. Not only does this reduce losses but with radioisotopes it also reduces the hazards.

In selecting the site for a label, the metabolism of the labelled substrate needs to be considered. A label in the carboxyl group of an amino acid may be suitable for some purposes in which the amino acid is incorporated as an intact unit. However, there are other situations such as the formation of a neurotransmitter from an amino acid where decarboxylation occurs and a carboxyl label would then be of no use. Amino acids can also undergo transamination to afford an $\alpha$-keto acid. Clearly, a tritium label at the $\alpha$-position would be unsatisfactory for these studies. In some biosyntheses (*e.g.* that of griseofulvin) there is a symmetrical intermediate and this can lead to the randomization of a label and the consequent difficulty in the interpretation of a labelling pattern. There can be randomization of acetate labels *via* the citric acid cycle. In selecting the site for a label, it is also worth remembering that an isotope effect may alter the distribution of a label between two competing pathways.

The chromatographic separation and identification of isotopically labelled compounds often rests on the assumption that the labelled compound and its authentic unlabelled standard co-chromatograph. A few cases have arisen where this is not the case and the isotopic fractionation of compounds has occurred. This occurs most often with tritium and a few examples have been noted where $^{13}H:^{14}C$ ratios vary with chromatographic fractions of the same compound.

## 1.5  THE DETECTION OF ISOTOPES

The unit of radioactivity is the becquerel (Bq), which is equivalent to one disintegration per second (dps). The older non-SI unit is the curie (Ci), which is equivalent to the number of disintegrations per second produced by one gram of radium. One curie corresponds to $3.7 \times 10^{10}$ dps or 37 GBq. For most preparative radiochemical labelling experiments described in this book, the becquerel is a rather small unit and the curie too large a unit. The more convenient units are the MBq or kBq ($10^7$ or $10^4$ dps) or mCi (1 mCi = 37 MBq) or μCi (1 μCi = 37 kBq). However, four other parameters are of major importance: the nature of the radiation (α-, β- or γ-radiation), the energy of the radiation (meV), the half-life of the isotope and the concentration of the radioactivity in the sample. The last of these parameters, known as the specific activity, is quoted in terms of MBq mmol$^{-1}$ or mCi mmol$^{-1}$. The nature of the radiation, the half-life of the isotope and the energy of the radiation will determine the method of measurement, the adsorbed dose and exposure or dose equivalent limits [units: gray (Gy) and sievert (Sv)], the toxicity classification of the isotope, the licensing conditions, the nature of the laboratory precautions and the standard of the laboratory fittings that are required. Most of the isotopic work that is described in this book concerns isotopes that fall into the low-toxicity (tritium) or medium-toxicity (carbon-14, sulfur-35 and phosphorus-32) groups.

With stable isotopes, the percentage enrichment at a particular site is usually recorded. The relationship of this to the natural abundance will determine the detection method and detection limits.

When biosynthetic experiments are carried out with either stable or radioactive isotopes, data such as the percentage incorporation of the label and its specificity may be reported. The dilution of the specific activity may also be a helpful parameter to consider in the light of the role of metabolic pools.

Up to the 1960s, the Geiger–Müller tube was the standard instrument for measuring radioactivity. It operated by detecting the ionization of a gas produced by the radiation. When these ions were produced in the gas between two oppositely charged electrodes, a discharge occurred which could be measured. These Geiger counters were often mounted within a 'lead castle' to shield them from external radiation. The radioactive sample would be counted as a solid derivative on a planchette. The efficiency depended on the energy of the radiation and was often low. Since the 1960s, this method has been superseded by liquid scintillation counting, although Geiger–Müller tubes are still widely used in monitoring equipment and in some radiochromatogram scanners.

Liquid scintillation counting is the preferred method for counting the commonly used weak β-emitters such as tritium and earbon-14. The less common isotopes such as phosphorus-32, sulfur-35 and chlorine-36 are also counted in this way. The radioactive sample is dissolved in a solvent such as toluene or xylene, sometimes with a solubilizing or emulsifying agent. The energy of each β-particle that is emitted in the radioactive decay process is transferred *via* the aromatic solvent to a fluor which emits a burst of light. This is detected by a photomultiplier tube and converted to a measurable electronic pulse. The fluors are typically 2,5-diphenyloxazole (PPO) and 1,4-bis(5-phenyloxazol-2-yl)benzene (POPOP). The primary scintillator PPO emits light at 365 nm, which is absorbed by the secondary scintillator (POPOP) and re-emitted at a longer wavelength (415 nm) close to the maximum efficiency of the photomultiplier tube. Under these conditions, counting efficiencies can be very high (~95%) for carbon-14 and good (~60%) for tritium.

Although the underlying principle is fairly simple, a number of precautions need to be taken. Two photomultipliers are operated with a coincidence circuit in order to discriminate between a genuine radiochemical event and a random thermal emission of an electron. The radioactive substance itself may absorb (quench) the light emission and a correction may have to be applied. Furthermore, the radioactive substance may be poorly soluble in the aromatic solvent. A second solvent such as 1,4-dioxane may be used, but this can reduce the counting efficiency. Various detergent additives such as Triton-XXX can be used for counting aqueous solutions. Additional counting channels allow the energy of the light pulse to be determined. In this way, tritium and carbon-14 may be determined in the same sample. One of the great advantages of liquid scintillation counting is that it can be automated to allow many samples in a batch to be counted without manual intervention. Aqueous solutions of higher energy β-emitters such as phosphorus-32 emit flashes of light as a result of radioactive decay (Cerenkov radiation) and these can be counted without the use of a liquid scintillant.

For many years, the detection of the site of a radioactive label in a metabolite involved a careful chemical degradation to produce fragments which uniquely identified a single site within the metabolite. In some cases this work also utilized mass spectrometric identification of the presence of stable isotopes such as deuterium or carbon-13 in the molecular and fragment ions. Useful information on the integrity of units and on the stereochemistry of biosynthetic events involving the displacement of hydrogen atoms came from the measurement of tritium:carbon-14 ratios and their changes between substrate and metabolite. The site of the label may therefore be determined by the ease with which these degradations can be undertaken.

Apart from early methods based on the combustion of a compound and determining the density of the water that was produced, up to the

1960s the presence of stable isotopes were almost entirely detected by mass spectrometry and their site located by the fragmentation pattern. Indeed, deuterium labelling played an important role in establishing the details of fragmentation patterns. Since that time, NMR methods have played an increasingly important role.

The detection of the site of a label by $^{13}$C NMR spectroscopy has had a major impact on biosynthetic studies. Carbon-13 has a natural abundance of 1.1% and therefore, provided that the signals in the $^{13}$C NMR spectrum have been assigned, enrichment above natural abundance at specific sites can be measured. Many biosynthetic problems have been solved because these enhancements can be measured on samples of a few milligrams.

More information on the integrity of units can be obtained from $^{13}$C–$^{13}$C couplings. At the natural abundance the statistical chance of having two carbon-13 nuclei adjacent to each other is too low for the $^{13}$C–$^{13}$C coupling to be easily observed. However, if a unit such as [$^{13}$C$_2$]acetate containing >90% of the label at both sites is fed to an organism which is biosynthesizing a metabolite derived from a polyketide chain of intact acetate units, the $^{13}$C NMR spectrum of the metabolite will reveal a $^{13}$C–$^{13}$C coupling pattern. Provided that the labelled acetate is diluted with sufficient unlabelled acetate so that in the resultant chain a labelled acetate unit is only adjacent to an unlabelled acetate, the couplings that are observed just reflect the integrity of the labelled unit. If the $^{13}$C–$^{13}$C bond has been broken during the biosynthesis, the coupling is lost. The overall coupling pattern which emerges is the summation of many chains in which the labelled acetate units have been randomly incorporated along the chain. It then becomes possible, using the coupling pattern, to identify the way in which a polyketide chain is folded to generate a metabolite. This use of $^{13}$C–$^{13}$C couplings can obviously be extended to other biosyntheses involving, for example, isoprenoid units. It is also possible to change the feeding protocol to identify couplings that may arise from bonds that are formed during a biosynthesis.

The carbon-13 chemical shift shows a small sensitivity to an isotope that is attached to it. Thus the chemical shift of a $^{13}$C–$^2$H or a $^{13}$C–$^{18}$O signal appears at a slightly different position from the corresponding $^{13}$C–$^1$H or $^{13}$C–$^{16}$O signal. This isotope shift of a carbon-13 resonance can be used as a reporter for the presence of a deuterium or oxygen-18 label. The effect is additive and hence the number of deuterium labels attached to a carbon-13 can be identified. An isotope shift is also observed on the adjacent β-carbon atom. The information that can be obtained from studying the fate of multiply labelled precursors has led to the development of strategies for the incorporation of several labels into molecules.

The sensitivity and resolution of high-field NMR spectrometers has made it possible to observe the spectra of peptides and other biological macromolecules. The chemical shifts of various nuclei are sensitive to intramolecular interactions arising from the folding of these chains. This in turn has provided the stimulus for the synthesis of specifically labelled amino acids and nucleic acid bases to identify these interactions.

A number of the other isotopes that are used to label organic compounds have a nuclear spin and can be detected by NMR spectroscopy. While the site and stereochemistry of a deuterium label in a highly deuteriated sample may be inferred from the changes to the $^1$H and $^{13}$C NMR spectra, it is also possible to record a $^2$H NMR spectrum. Deuterium has a spin of 1. In a magnetic field of 2.35 T when the proton resonates at 100 MHz, deuterium spectra are recorded at 15.35 MHz. The sensitivity is fairly low and the signal is broadened by quadrupolar relaxation. Spectra are usually observed with proton noise decoupling. Although in frequency terms (Hz) the spectra are spread over a rather narrow range, it is possible to establish sites of deuteriation from the chemical shift by comparison with the proton spectrum.

Tritium has a nuclear spin of 1/2 and, when the proton is observed at 100 MHz, the tritium resonances are at 106.5 MHz. Consequently, in frequency terms the spectra are spread over a wider range than with deuterium and with proton noise decoupling sharp signals can be observed. Although $^3$H NMR detection is fairly sensitive, even with a micro-cell, it has hitherto required the equivalent of 1 mCi of radioactivity at a site for a measurement to be made within a reasonable timescale on a modern Fourier transform instrument. Hence this has placed many, but not all, biosynthetic applications outside the realms of $^3$H NMR. However, sensitivity improvements involving novel probes may significantly reduce the amount of tritium that is needed. Nevertheless, the method has been useful in determining the specificity of labelling by chemical means. It has been particularly useful in determining the distribution of the label in compounds that have been generally labelled by the Wilzbach method. In these examples the tritium may be unevenly distributed over a number of centres. It has also been used to establish the extent and stereochemistry of labelling when tritium has been introduced catalytically. However, the potential radiochemical hazard associated with these measurements has restricted their use.

The common isotope of nitrogen ($^{14}$N, natural abundance 99.6%) has a spin of 1 whereas nitrogen-15 has a spin of $\frac{1}{2}$. Both have a low magnetic moment and a low NMR frequency (at 2.35 T, $^{14}$N

7.2 MHz, $^{15}$N 10.1 MHz). Because of quadrupolar relaxation, the $^{14}$N NMR signals are often very broad and hence more attention has been devoted to the measurement of the $^{15}$N NMR spectra of enriched samples. The nitrogen-15 resonances cover a wide range of chemical shifts. Useful correlations exist between both the chemical shift and coupling constants and chemical structure.

Oxygen-17 has a spin of 5/2 and at 2.35 T spectra are recorded at 13.5 MHz. Again, because of the quadrupole moment, the linewidths are broad and the sensitivity is low compared with the proton. Nevertheless, correlations exist between chemical shift and structure.

A wide range of physical methods exist for the detection of isotopes. The success of an isotopic method in studying a chemical or biochemical problem can depend heavily on the selection of an appropriate method for detecting the label.

## 1.6   NOMENCLATURE

The IUPAC rules for the naming of isotopically modified compounds form Section H of the IUPAC Rules on the Nomenclature of Organic Compounds. When an isotope is being used in modifying a compound, this is described using the atomic symbol of the nuclide with the isotope being shown as a left-hand superscript as in $^2$H or $^{13}$C. The IUPAC rules distinguish between several different isotopic modifications to a compound. Thus a distinction is made between 'isotopic substitution' in which all the molecules of the compound have only the indicated nuclide at the particular site and a 'labelled compound' in which there is just an un-natural proportion of the nuclide at a particular site. In practice, except for the introduction of deuterium and some compounds containing carbon-13, the majority of the compounds described in this book fall into the latter group. Where isotopic substitution has occurred, the modification is represented by placing the isotopic substituent in parentheses as in (1-$^2$H$_1$)ethanol. However, when isotopic labelling is involved, square brackets are used as in [1-$^2$H$_1$]ethanol.

The rules also distinguish between a generally labelled compound and a uniformly labelled compound. A generally labelled compound will have the label unevenly distributed throughout the compound whereas a uniformly labelled compound will have an even distribution. Tritium introduced by a Wilzbach exchange will afford a generally labelled compound whereas a compound biosynthesized in an atmosphere of carbon-14-labelled carbon dioxide as the sole source of carbon may be uniformly labelled.

When it comes to defining the site of the label, the position and nuclide symbol are placed in the relevant brackets preferably immediately before the name of the part of the compound in which the label is located. A typical example is ethyl [2-$^{13}$C]acetate. Note that there is no hyphen between the bracket and the name. Where there is multiple-labelling, then both positions are numbered, as in ethyl [1,2-$^{13}$C$_2$]acetate. If two different isotopes are involved, they are cited in alphabetical order, as in ethyl [2-$^{13}$C,2-$^2$H$_2$]acetate. Where the isotopes are of the same element as in deuterium and tritium, the order is that of the mass number. Where a chiral designation, (*R*) or (*S*) (see Chapter 4), is involved, this precedes the isotopic designation, as for example in (*R*)-[2-H$_1$,$^2$H$_1$,$^3$H$_1$]acetic acid. Where there is a choice in numbering a hydrocarbon chain, the starting point that is chosen is the one that assigns the lowest number to the site of the isotopic label. Where there are two otherwise identical chains, the group bearing the label should be incorporated into the stem and not treated as a substituent. Sometimes for the sake of clarity, particularly with natural products, the group bearing the label may be specified in full, for example L-[*methyl*-$^{13}$C]-methionine. The IUPAC rules cite many other useful examples.

There are some other systems which have been used, particularly in the biochemical literature. In one system deuterium is designated as *d* and tritium as *t*. For example, a labelled form of methionine might be described as L-[*methyl*-*d*$_3$]methionine. However, the site of the label is usually apparent from the context. A clear diagram including structural formulae with the sites of labelling is particularly useful.

CHAPTER 2

# Labelling Compounds with Carbon-13 and Carbon-14

## 2.1 INTRODUCTION

The radioactive isotopes of carbon, carbon-11 and carbon-14, and the stable isotope, carbon-13, have played a very important role in many chemical and biochemical studies. Although carbon-11 was used in some of the earliest studies with carbon-labelled compounds, its short half-life ($t_{\frac{1}{2}} \approx 20$ min) and the increasing availability of carbon-13 and carbon-14 precluded its widespread use until there was a resurgence of interest associated with the advent of positron emission tomography (PET). This short half-life imposes special constraints on the synthesis of carbon-11-labelled compounds and consequently these syntheses will be discussed later in Chapter 9 on the use of isotopes for PET. Nevertheless, many of the recent organometallic methods which have been used in the context of carbon-11 labelling have the potential for use with carbon-13 or carbon-14.

Carbon-13 is a stable isotope of carbon which is present in 1.1% natural abundance. There are tiny variations in the carbon-12:carbon-13 ratio in some natural products such as sugars and the compounds formed from them. These variations reflect isotope effects in the different pathways for the biosynthesis of sugars. These isotope ratios can have diagnostic applications in, for example, forensic analysis.

The main method for the separation of carbon-13 is based on the cryogenic distillation of carbon monoxide at −190 °C. [$^{13}$C]Carbon monoxide of

The Organic Chemistry of Isotopic Labelling
By James R. Hanson
© James R. Hanson 2011
Published by the Royal Society of Chemistry, www.rsc.org

95–99% isotopic purity can be obtained by this means. This process also produces [$^{12}$C]carbon monoxide depleted in carbon-13. The carbon monoxide is either oxidized to carbon dioxide or reduced to methanol from which methyl iodide and the starting materials for other syntheses can be obtained. Sodium [$^{13}$C]cyanide is also produced *via* cyanamide.

Carbon-14 is obtained in a nuclear reactor by the nuclear reaction

$$^{14}N + n \rightarrow {}^{14}C + p$$

The nitrogen is held as beryllium or aluminium nitride. There is usually sufficient oxygen present to facilitate the conversion of the carbon-14 to [$^{14}$C]carbon dioxide, which becomes the primary source of the label. This can be readily converted to methanol, simple carboxyl-labelled acids, methyl iodide, potassium cyanide, formic acid and other one-carbon units which form the starting materials for synthesis. The majority of the methods that are important for the introduction of a carbon label into an organic compound are based on activating single- or two-carbon units.

Although the precautions, the scale and the methods of detection differ between carbon-13 and carbon-14, many of the reactions that are used to introduce these labels into organic molecules are the same. Hence they will be considered together.

## 2.2 CARBON LABELLING USING THE CYANIDE ION

The cyanide anion is a very useful one-carbon unit for introducing a label into a compound. The cyanide ion can participate in nucleophilic substitution and addition reactions. An aryl cyanide can be prepared from the cyanide ion and a diazonium salt. Once the labelled cyano group has been introduced into a molecule, it can undergo further transformations. Thus reduction leads to primary amines, hydrolysis proceeds *via* the amide to a carboxylic acid and other carbon and nitrogen nucleophiles can add to the electron-deficient carbon of the cyano group leading to, for example, the synthesis of heterocyclic compounds. The acid-catalysed Ritter reaction of a cyanide with the carbocation derived from an alcohol is a useful way of preparing an amide. A formamide is produced from hydrogen cyanide by this reaction. A cyano group will render an adjacent C–H weakly acidic and it will participate in the stabilization of the resultant carbanion. Consequently, it is not surprising that the cyanide ion has played an important role in labelling organic compounds, particularly by cyanohydrin formation from carbonyl groups and by the nucleophilic substitution of alkyl halides.

The addition of hydrogen [$^{14}$C]cyanide to acetaldehyde and hydrolysis of the cyano group affords the α-hydroxy acid [$^{14}$C]lactic acid (**2.1**). Labelled glycerol **2.4** has been prepared (Scheme 2.1) by protecting the primary alcohol of the C$_2$ hydroxy aldehyde, glyoxal as its benzyl ether **2.2**. The aldehyde was then converted to the cyanohydrin using potassium [$^{14}$C]cyanide. The cyano group was hydrolysed and the corresponding ester **2.3** was reduced with lithium aluminium hydride. Finally deprotection gave [1-$^{14}$C]glycerol (**2.4**).

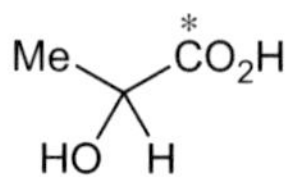

**2.1**

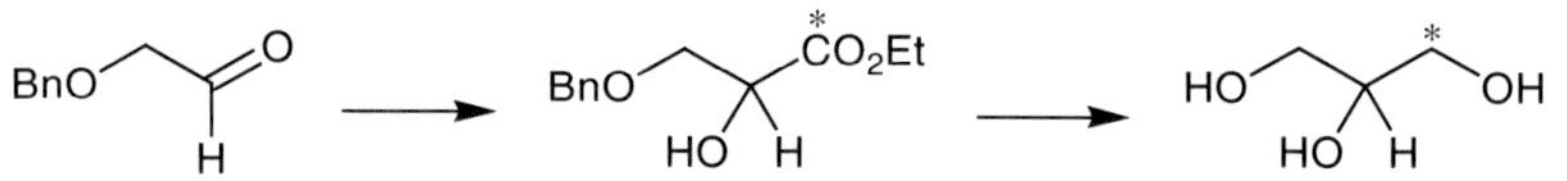

| 2.2 | 2.3 | 2.4 |

**Scheme 2.1**   Synthesis of [$^{14}$C]glycerol.

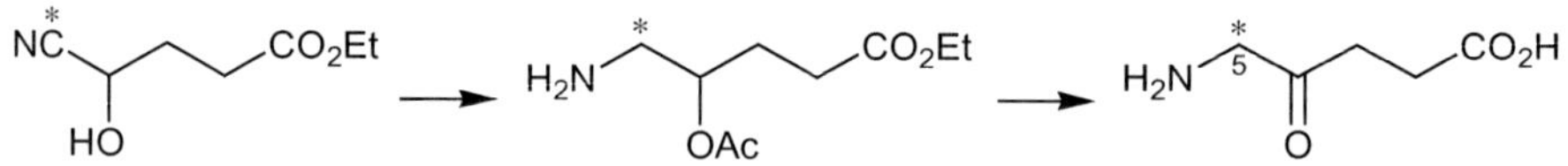

| 2.5 | 2.6 | 2.7 |

**Scheme 2.2**   Preparation of labelled δ-aminolaevulinic acid.

The porphyrin macrocycles which are constituents of vitamin B$_{12}$ and the haem pigments are built up from the pyrrole porphyrobilinogen, which is in turn assembled from δ-aminolaevulinic acid (**2.7**). Studies on the biosynthesis of the haem ring system have required [$^{13}$C]- and [$^{14}$C]-labelled samples of δ-aminolaevulinic acid. A straightforward synthesis used labelled sodium cyanide (Scheme 2.2). The aldehyde of ethyl 4-oxobutyrate was converted to the cyanohydrin **2.5**. The hydroxyl group was then acetylated and the cyano group was hydrogenated in acidic ethanol to give 4-acetoxy-5-aminopentanoic acid (**2.6**). Hydrolysis and oxidation of the hydroxyl group under mild conditions gave labelled δ-aminolaevulinic acid (**2.7**) labelled at C-5.

The nucleophilic displacement of a halogen by the cyanide ion and the subsequent hydrolysis of the cyano group to an acid has been used to label a number of carboxylic acids. For example, treatment of 1,5-dibromopentane with potassium [$^{14}$C]cyanide and hydrolysis of the product gave the $C_7$ acid, [1,7-$^{14}$C$_2$]pimelic acid. Pyrolysis of the calcium salt of this dicarboxylic acid gave [1-$^{14}$C]cyclohexanone. This labelled cyclohexanone has been used to establish a cyclopropanone mechanism for the Favorskii rearrangement of α-halo ketones to carboxylic acids.

Another example which illustrates several of these reactions is the preparation of [1,5-$^{14}$C$_2$]citric acid (**2.9**) (Scheme 2.3). Addition of hydrogen cyanide to 1,3-dichloroacetone afforded a cyanohydrin, which on hydrolysis with hydrochloric acid gave the corresponding di(chloromethyl)-α-hydroxy acid **2.8**. The label was introduced by nucleophilic substitution of the chlorine atoms with potassium [$^{14}$C]cyanide. Hydrolysis afforded the labelled citric acid **2.9**. Examination of this scheme reveals its potential application for introducing labels at other sites.

When the Krebs citric acid cycle (the tricarboxylic acid cycle) was first described, a problem was raised because one of the intermediates, α-ketoglutaric acid (2-oxopentanedioic acid), was unsymmetrically labelled by sodium acetate that had been introduced into the biosynthetic system. This raised the question of how an apparently symmetrical intermediate, citric acid, might be involved in the cycle which had an apparent overall asymmetry. Ogston resolved this difficulty by introducing the concept of prochirality. He pointed out that if there was a three-point contact between citric acid and an asymmetric enzyme, the asymmetry of the enzyme–substrate complex would lead to a distinction between the –CH$_2$CO$_2$H arms of the citric acid. This would be reflected in the conversion of the enzyme-bound citric acid to *cis*-aconitic acid (*cis*-propene-1,2,3-tricarboxylic acid) and thence to α-ketoglutaric acid. A proof of this required a sample of citric acid asymmetrically labelled in one arm (Scheme 2.3). γ-Chloro-β-carboxy-β-hydroxybutyric acid (**2.10**) was prepared by a cyanohydrin reaction of ethyl 4-chloroacetoacetate (ClCH$_2$COCH$_2$CO$_2$Et) followed by hydrolysis. The acid was then resolved into its enantiomers using its brucine salt. The laevorotatory enantiomer was treated with sodium [$^{14}$C]cyanide and the resultant nitrile hydrolysed to give asymmetrically labelled citric acid. When this enantiospecifically labelled citric acid was subjected to

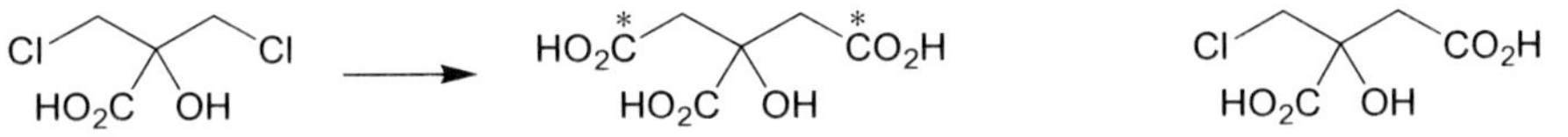

**Scheme 2.3** Synthesis of [$^{14}$C]citric acid.

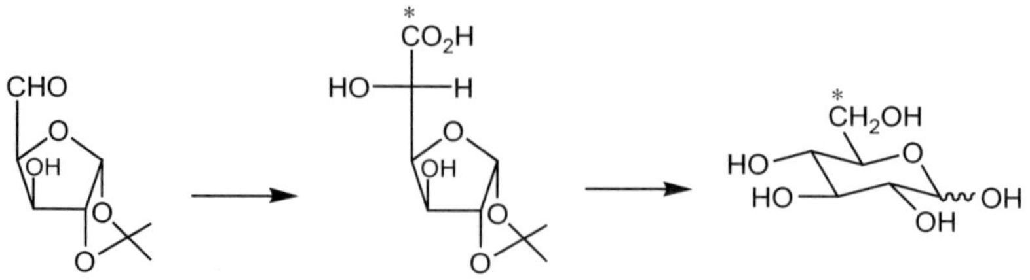

**Scheme 2.4**   Synthesis of [6-$^{14}$C]glucose.

enzymatic action, the resultant α-ketoglutaric acid arising from this enantiomer was found to carry the isotope entirely in the γ-carboxyl group.

The preparation of D-[6-$^{14}$C]glucose (**2.13**) (Scheme 2.4) represents a further example of cyanohydrin formation. In this example, the retention of the stereochemistry of the sugar was important. This particular labelled glucose is of interest because various biochemical processes involve the scission of the carbon chain of D-glucose. The procedure started with D-glucose and aimed to retain as far as possible the stereochemical integrity of the sugar. The terminal carbon of D-glucose was removed as form-aldehyde by the cleavage of 1,2-isopropylidene-D-glucofuranose with sodium periodate. The resultant C$_5$ aldehyde **2.11** was then treated with sodium [$^{14}$C]cyanide to reconstitute the C$_6$ carbon skeleton. Hydrolysis of the cyanohydrin gave a separable mixture of the epimeric hydroxy acids [6-$^{14}$C]idouronic and [6-$^{14}$C]glucuronic acids (**2.12**). The latter formed a γ-lactone, which on reduction with sodium borohydride and hydrolysis of the acetonide gave [6-$^{14}$C]glucose (**2.13**).

[$^{14}$C]Benzonitriles and the corresponding aromatic carboxylic acids have been prepared by a palladium-catalysed cyanation of aryl iodides. Thus an aryl iodide such as 2-iodoaniline was treated with zinc [$^{14}$C]cyanide and palladium tetrakistriphenylphosphine in dimethylformamide to give 2-amino[7-$^{14}$C]benzonitrile. Basic hydrolysis of the cyano group afforded [7-$^{14}$C]anthranilic acid. An alternative method is to treat 2-iodoaniline with potassium [$^{14}$C]cyanide and copper(I) iodide in hot dimethylforma-mide. Labelled heterocyclic compounds such as 4-aminoquinazolines **2.14** have been prepared from the 2-amino[7-$^{14}$C]benzonitrile.

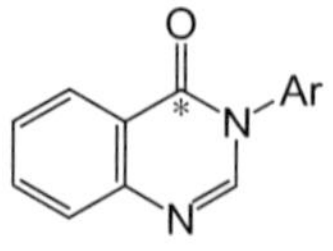

**2.14**

## 2.3 ORGANOMETALLIC REAGENTS IN CARBON LABELLING

Organometallic compounds provide carbanion reagents which have played a significant role in the synthesis of labelled compounds. The preparation of these from alkyl halides results in the reversal of the polarity of the carbon–halogen bond and a change in the reactive character of this carbon atom from being a site of nucleophilic attack to a centre which behaves as a nucleophile. The reactivity of the organometallic carbanion varies with the associated metal (commonly Li, Mg, Cu, Cd or Zn). Although the majority of the reactions are of a carbanion character, some organometallic reagents may act as free radicals. The metal ion may coordinate to the oxygen of a carbonyl group or an epoxide increasing the electron deficiency of an attached carbon atom and facilitating the attack by a carbanion. Metal salts may be added to the organometallic reagent for this specific purpose.

The Grignard reagents are the best known of these compounds. They are prepared by adding magnesium to a solution of an alkyl halide in diethyl ether. The magnesium inserts into the carbon–halogen bond. The diethyl ether coordinates to the magnesium of the Grignard reagent. Provided that there is sufficient ether present to fulfil this stabilizing role, a higher boiling solvent such as toluene may be added so that a difficult reaction may be heated.

The major single-carbon units that are widely employed in organometallic labelling reactions are the Grignard reagent prepared from [$^{13}$C or $^{14}$C]methyl iodide and the use of [$^{13}$C or $^{14}$C]carbon dioxide as a substrate for a Grignard reaction. It is also worth noting that 'protonolysis' of a Grignard reagent with deuterium oxide or tritiated water provides a method for introducing deuterium or tritium into a molecule (see Chapter 3).

The reaction of labelled [$^{2}$H, $^{3}$H, $^{13}$C or $^{14}$C]methylmagnesium iodide with carbon dioxide provides the main method of preparing methyl-labelled acetic acid. If carbon-13- or carbon-14-labelled carbon dioxide is used, the result is a carboxyl-labelled acetic acid. Simple labelled building blocks can then be synthesized. These include the amino acid glycine ($H_2NCH_2CO_2H$), which is obtained by bromination and amination. Labelled acetone (propan-2-one) is obtained by pyrolysis of the calcium salt of acetic acid. Cyanoacetic acid and malonic acid ($HO_2CCH_2CO_2H$) can be obtained by substitution of the bromoacetic acid with cyanide, and ethyl acetoacetate ($CH_3COCH_2CO_2Et$) can be obtained by a Claisen condensation of ethyl acetate.

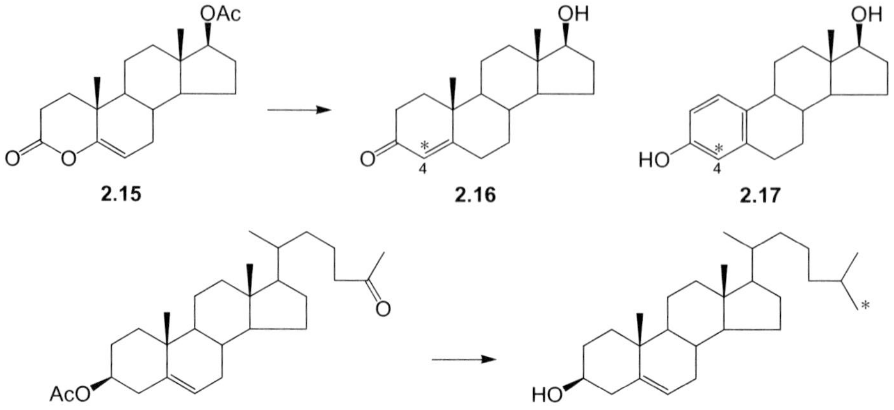

**Scheme 2.5**   Preparation of some carbon-labelled steroids.

Pyruvic acid (CH₃COCO₂H) is a central metabolic intermediate and various labelled forms have been prepared to establish the role that it plays in metabolism. [3-¹⁴C]Pyruvic acid has been prepared by the reaction of [¹⁴C]methylmagnesium iodide with diethyl oxalate [(CO₂Et)₂].

A simple Grignard reaction with labelled methylmagnesium iodide can provide a method for introducing a label into complex molecules such as the steroids (Scheme 2.5). The hormone testosterone (**2.16**) has been labelled at C-4 by the oxidative cleavage of the unsaturated ketone on ring A to form a 4-nor-5-keto acid. This was converted to the enollactone **2.15**, which on reaction with [¹⁴C]methylmagnesium iodide gave [4-¹⁴C]testosterone (**2.16**). Modification of the procedure allowed it to be used with the other hormonal steroids. Thus application of the sequence to 19-nortestosterone followed by a dehydrogenation gave [4-¹⁴C]estradiol (**2.17**).

The side-chain of cholesterol has been labelled using a Grignard reaction. The 27-nor-25-oxocholesterol acetate **2.18** has been obtained by debromination of one of the products of oxidation of the dibromide of cholesteryl acetate. A Grignard reaction with [¹⁴C]methylmagnesium iodide gave labelled 25-hydroxycholesterol acetate. [26-¹⁴C]Cholesterol (**2.19**) was obtained by dehydration, careful partial hydrogenation and hydrolysis.

Grignard procedures have been used to obtain carboxyl-labelled fatty acids. Thus [1-¹⁴C]palmitic acid has been prepared by a Grignard reaction of the corresponding nor-bromide with [¹⁴C]carbon dioxide. The nor-bromo compound may be prepared by the bromodecarboxylation of the silver salt of the carboxylic acid (the Hunsdiecker reaction).

[1,4-$^{14}$C$_2$]Succinic acid (HO$_2$CCH$_2$CH$_2$CO$_2$H) has been prepared by carboxylation of the dilithium salt derived from acetylene with [$^{14}$C]carbon dioxide. Hydrogenation of the labelled acetylenedicarboxylic acid gave [1,4-$^{14}$C$_2$]succinic acid.

## 2.4 THE WITTIG REACTION IN CARBON LABELLING

The Wittig reaction provides a regiospecific method for creating an alkene in place of a carbonyl group. Labelled methyltriphenylphosphonium iodide can be easily prepared from labelled methyl iodide and triphenylphosphine and hence there is a convenient means of introducing a carbon-13 or carbon-14 label into a molecule *via* the corresponding ylid, which is formed by the action of strong base. The reaction of a ketone with the nucleophilic carbon of the methylenetriphenylphosphorane affords a betaine, which undergoes a thermal decomposition to eliminate triphenylphosphine oxide and regiospecifically form the labelled alkene.

An alkene which contains a terminal methylene can be a suitable site for labelling by a Wittig reaction. Isopentenyl diphosphate is a key intermediate in the biosynthesis of isoprenoid compounds. The parent alcohol, isopentenol (**2.20**), was prepared in a labelled form by a Wittig reaction between 4-acetoxybutan-2-one (CH$_3$COCH$_2$CH$_2$OAc) and [$^{14}$C]methylenetriphenylphosphorane followed by hydrolysis of the acetate. Where there is an alkene in the target molecule, it may be cleaved by ozonolysis or potassium permanganate–sodium periodate oxidation to afford a nor-ketone. The alkene may then be reinstated by a Wittig reaction with a labelled ylid. The hydrocarbon *ent*-kaurene (**2.21**) co-occurs with the plant hormone gibberellic acid in the fungus *Gibberella fujikuroi*. Its role in the biosynthesis of the gibberellins was established using labelled material obtained by oxidizing the *ent*-kaurene to its 17-nor-16-ketone and then restoring the 16-alkene with the ylid from [$^{14}$C]methyltriphenylphosphonium iodide to give *ent*-[17-$^{14}$C] kaurene (**2.21**).

Abscisic acid (**2.23**) is a plant hormone which is involved in the control of dormancy, abscission (leaf fall) and the response of the plant to water stress. Labelled abscisic acid was required for metabolic studies and as an internal standard for use in gas chromatographic–mass spectrometric analysis. [1,2-$^{13}$C$_2$]Carbomethoxymethylenetriphenylphosphorane was prepared from the methyl ester of [1,2-$^{13}$C$_2$]-bromoacetic acid (Scheme 2.6) and used in a Wittig reaction with 1-hydroxy-4-keto-α-ionone (**2.22**) to give a mixture of the *cis* and *trans* isomers of the methyl ester of [1,2-$^{13}$C$_2$]abscisic acid (**2.23**). The methyl esters were hydrolysed and the acids were separated by chromatography. This procedure reveals the lack

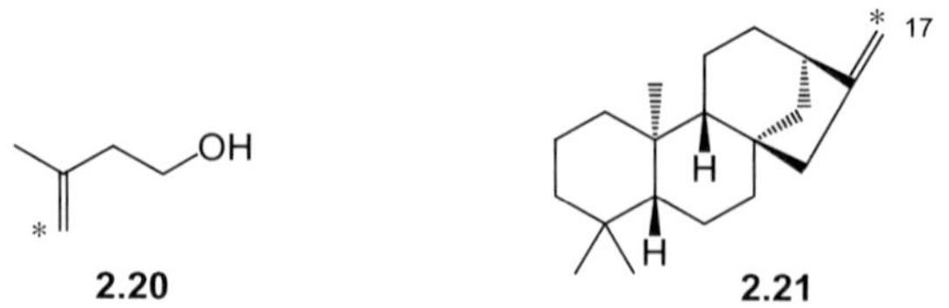

**Scheme 2.6** Some applications of the Wittig reaction in labelling.

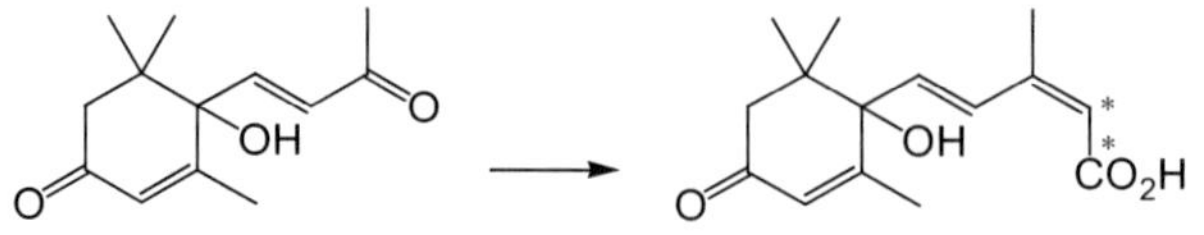

of stereospecificity of the Wittig reaction and the tendency for it to give a significant proportion of the *cis* isomer. In this particular instance, the formation of the *cis* isomer was useful since abscisic acid possesses the *cis* geometry. Similar procedures have been used in labelling compounds in the vitamin A and retinoic acid series.

The Wadsworth–Emmons phosphonate modification of the Wittig reaction overcomes this lack of stereospecificity. In this case, a carbanion is generated from a phosphonoacetate (*cf.* Section 2.5). The first steps in the reaction of the phosphonate carbanion with a carbonyl group are reversible and hence there is a tendency for the reaction to give adducts which lead to the less hindered *trans*-alkenes. This has been particularly useful in the synthesis of labelled terpenoid precursors based on geranyl, farnesyl and geranylgeranyl diphosphates (Scheme 2.7). Thus treatment of methyl [1-$^{14}$C]bromoacetate with trimethyl phosphite gave the phosphonate **2.24**. Reaction of the ylid from this phosphonate with geranylacetone (**2.25**) gave methyl [1-$^{14}$C]farnesoate, from which [1-$^{14}$C]farnesol (**2.26**) was obtained by reduction with lithium aluminium hydride. The farnesol was then diphosphorylated.

The [1,2-$^{13}$C$_2$]phosphonate equivalent to **2.24** can be prepared from methyl [1,2-$^{13}$C$_2$]bromoacetate. This provides a very useful way of introducing a $^{13}$C$_2$ unit into a molecule. The retention of a $^{13}$C–$^{13}$C coupling in a biosynthetic sequence is a useful indicator of the integrity of a bond in that sequence. [1,2-$^{13}$C$_2$]-5,5-Dichlorohexanoic acid (**2.28**), which was required for biosynthetic studies, was prepared from the protected butanal **2.27** and triethyl [1,2-$^{13}$C$_2$]phosphonoacetate. The double bond of the unsaturated ester was reduced and the protecting group removed to give ethyl [1,2-$^{13}$C$_2$]-5-oxohexanoate, from which the geminal dichloride was obtained.

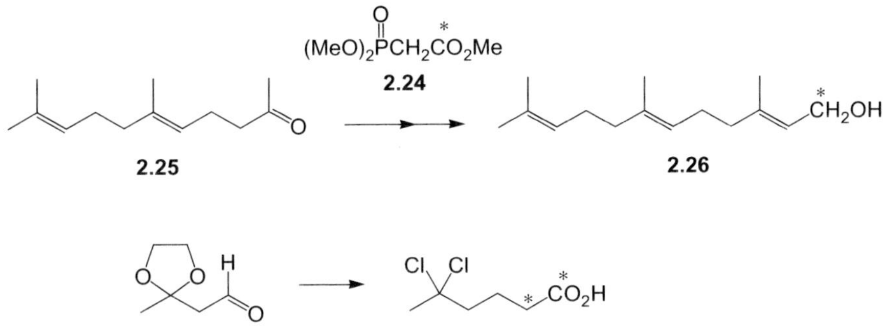

**Scheme 2.7**   Some applications of the Wadsworth–Emmons reaction in labelling.

## 2.5   CARBONYL CONDENSATION REACTIONS IN CARBON LABELLING

The presence of a carbonyl group can make a hydrogen atom attached to an adjacent atom acidic. The reaction of this hydrogen atom with a base can then generate an anion which achieves stabilization by de-localization over the carbonyl group. In the case of a C–H, the anion is a carbanion and resonance stabilization is provided by the involvement of an enolate anion. The carbanion is a useful nucleophile which can participate in substitution and addition reactions leading to the for-mation of new carbon–carbon bonds. Reactions of this type play a significant part in the introduction of carbon labels into molecules. The presence of a second carbonyl group adjacent to a C–H, as in a 1,3-diketone, increases the acidity of the C–H. This activating effect is not restricted to a carbonyl group but is also displayed by nitro, phospho-nate, sulfoxide and sulfono groups. An example has already been seen in the Wadsworth–Emmons modification of the Wittig reaction.

Malonic acid and its esters and ethyl acetoacetate are the commonest building blocks that utilize this chemistry. We have already seen that malonic acid in a labelled form can be prepared *via* cyanoacetic acid. Another route (Scheme 2.8) that has been used to prepare diethyl [2-$^{14}$C]malonate (**2.30**) involved the condensation of ethyl [2-$^{14}$C]acetate with diethyl oxalate to give diethyl oxaloacetate (**2.29**). On heating, this underwent decarbonylation to form diethyl [2-$^{14}$C]malonate (**2.30**).

Oxaloacetic acid is an intermediate in the Krebs citric acid (tri-carboxylic acid) cycle. When the acid is required in a labelled form it is better prepared by this route using the *tert*-butyl esters of acetic and oxalic acids. The di-*tert*-butyl ester of oxaloacetic acid can be hydrolysed

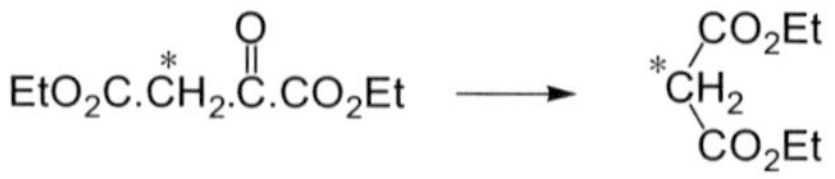

$$\underset{\textbf{2.29}}{EtO_2C.\overset{*}{C}H_2.\overset{O}{\overset{||}{C}}.CO_2Et} \longrightarrow \underset{\textbf{2.30}}{\overset{*}{C}H_2\underset{CO_2Et}{\overset{CO_2Et}{\big<}}}$$

**Scheme 2.8**   Synthesis of diethyl [2-$^{14}$C]malonate.

under mildly acidic conditions. The diethyl oxaloacetate also provided a source of other labelled C$_4$ compounds. Thus the keto group could be reduced to give the alcohol, diethyl malate, which in turn underwent dehydration to diethyl fumarate (diethyl *trans*-ethene-1,2-dicarboxylate). Hydrogenation of diethyl fumarate gave diethyl succinate. The parent acids were obtained by hydrolysis.

An alternative synthesis of [2-$^{14}$C]succinic acid to that described earlier proceeded from ethyl chloroacetate. The chlorine was displaced by the carbanion derived from diethyl [2-$^{14}$C]malonate (**2.30**). Hydrolysis of the product and decarboxylation of the malonic acid gave the labelled succinic acid. The same reaction sequence has been used to prepare [2-$^2$H$_2$]succinic acid.

Another high-yielding route which has been used to convert ethyl [2-$^{13}$C]acetate into diethyl [2-$^{13}$C]malonate is to deprotonate the ethyl acetate with the strong base lithium hexamethyldisilazide and then treat the carbanion with ethyl chloroformate.

Glycerol (**2.4**) labelled on either C-1 or C-2 has been obtained by oxidizing diethyl malonate with lead tetraacetate to give diethyl acetoxymalonate. Reduction of this compound with lithium aluminium hydride gave the triol glycerol.

Ethyl acetoacetate is conveniently prepared by a Claisen condensation of ethyl acetate, but there have been occasions when a more specifically labelled isotopomer such as ethyl [3-$^{14}$C]acetoacetate (**2.32**) was required. This was prepared by acylation of the mixed ester *tert*-butyl ethyl malonate (**2.31**) with [1-$^{14}$C]acetyl chloride (Scheme 2.9). *tert*-Butyl esters can be hydrolysed by acid with the elimination of isobutene. In this case, the *tert*-butyl ester was hydrolysed with acid and the subsequent decarboxylation of the monocarboxylic acid gave ethyl [3-$^{14}$C]acetoacetate (**2.32**). Alkylation of this labelled ethyl acetoacetate with geranyl bromide gave the labelled geranylacetone **2.25**, which was used in a synthesis of labelled squalene.

The nitro group is a useful activating group for the generation of carbanions. Once the nitro group has served its purpose as an activating functional group, it may be modified by reduction to a primary amine or converted to a carbonyl group. Nitromethane is a useful one-carbon

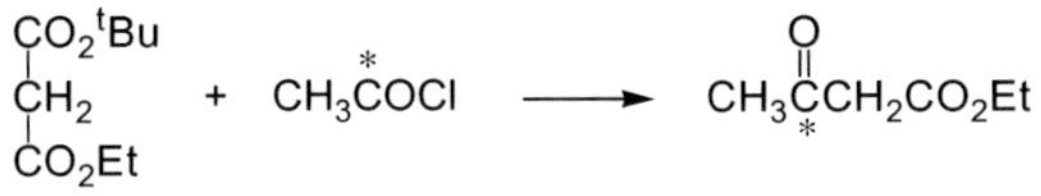

Scheme 2.9 structure:

CO₂ᵗBu / CH₂ / CO₂Et  (**2.31**)  +  CH₃*COCl  →  CH₃C(=O)CH₂*CO₂Et  (**2.32**)

**Scheme 2.9**  Synthesis of ethyl [3-$^{14}$C]acetoacetate.

Scheme 2.10 structures:

*CH₃NO₂ (**2.33**) → HOCH₂–*C(H)(NO₂)–HOCH₂ (**2.34**) → HOCH₂–*C=O–HOCH₂ (**2.35**) → CH₂OH / *CHOH / CH₂OH (**2.4**)

**Scheme 2.10**  An application of [$^{14}$C]nitromethane.

unit which can be used to introduce a carbon label into a molecule by such a carbanion process. [$^{14}$C]Nitromethane (**2.33**) may be prepared from [$^{14}$C]methyl iodide by treatment with silver nitrite. The preparation of [2-$^{14}$C]dihydroxyacetone (**2.35**) and [2-$^{14}$C]glycerol (**2.4**) exemplifies its application (Scheme 2.10). After the condensation of [$^{14}$C]nitromethane (**2.33**) with formaldehyde to give 1,3-dihydroxy-2-nitropropane (**2.34**), the nitro group was converted to a carbonyl group by the Nef reaction to produce 1,3-dihydroxyacetone (**2.35**) and thence, by reduction, [2-$^{14}$C]glycerol (**2.4**). A further application of a condensation with [$^{13}$C]nitromethane can be seen later (Scheme 2.14) in the preparation of 4-nitro[4-$^{13}$C]anisole.

## 2.6  THE SYNTHESIS OF CARBON-LABELLED MEVALONIC ACID

A number of the strategies described in the previous sections are illustrated by some syntheses of carbon-labelled samples of mevalonic acid (**2.36**) (Scheme 2.11). This compound is a key intermediate in the biosynthesis of the terpenoids and steroids particularly in mammals and fungi. Decarboxylation of the 3*R*-enantiomer of the 5-diphosphate of mevalonic acid affords isopentenyl diphosphate, which provides the characteristic isoprene building block of these natural products. Variously labelled forms of mevalonic acid have been used in studies of isoprenoid biosynthesis. Many of the cyclizations of the polyisoprenoid chains that lead to the individual families of terpenoids are accompanied by skeletal rearrangements. Different carbon-labelled forms of mevalonic acid have been required in order to elucidate the origin of the

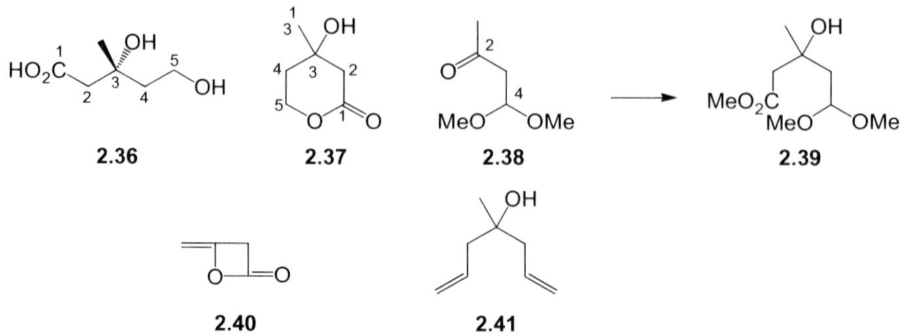

**Scheme 2.11**  Synthesis of carbon-labelled mevalonic acid.

various carbon atoms in these natural products. Although the bio-synthetically active form of mevalonic acid is the free carboxylic acid, it is normally-prepared as the cyclic δ-lactone **2.37**.

Retrosynthetic analysis of the structure of mevalonolactone in terms of labelled building blocks reveals potentially useful synthetic strategies which exemplify the use of carbonyl chemistry in carbon labelling. The presence of a β-hydroxycarbonyl moiety implies that a condensation reaction between a $C_2$ acetate unit and a 4-substituted butan-2-one would lead to the carbon skeleton. The presence of a tertiary alcohol at C-3 also suggests that other routes based on Grignard reactions might be useful. Another feature that has been used in the syntheses is to interchange the functionality on the two $C_2$ arms of mevalonolactone, converting the C-1 carboxyl to the C-5 primary alcohol and *vice versa*. Thus labels which were originally introduced at C-1 or C-2 then appear at C-4 or C-5.

The first samples of ( ± )-[2-$^{14}$C]mevalonolactone (**2.37**) were prepared by a Reformatsky reaction between methyl [2-$^{14}$C]bromoacetate and 4-acetoxybutan-2-one. This method was also used to prepare [4-$^{14}$C]mevalonolactone by carrying out the condensation with 4,4-dimethoxybutan-2-one (**2.38**). The ester in the product **2.39** was reduced with lithium aluminium hydride to a primary alcohol and the dime-thoxyacetal was then hydrolysed to an aldehyde. Mild oxidation of the aldehyde then afforded the [4-$^{14}$C]mevalonolactone **2.37**. The hydrolysis and oxidation have been carried out with peracetic or performic acid or with bromine water. Although a Reformatsky reaction was originally used for the addition of the $C_2$ unit to a butanone, the condensation can be achieved with a lithioacetate carbanion.

The $C_4$ butan-2-one unit has been prepared labelled at various centres in order to prepare different mevalonates for particular studies.

[3-$^{13}$C]Mevalonolactone was prepared *via* 4,4-dimethoxy[2-$^{13}$C]butan-2-one (**2.38**), which was obtained from the condensation of [2-$^{13}$C]acetone with methyl formate in the presence methanolic sodium methoxide followed by treatment of the hydroxymethylene compound with methanolic hydrogen chloride. [5-$^{13}$C]Mevalonolactone was prepared in a similar manner *via* 4,4-dimethoxy[4-$^{13}$C]butan-2-one (**2.38**), which was obtained from the condensation of acetone with methyl [$^{13}$C]formate.

Doubly labelled [3',4-$^{13}$C$_2$]mevalonolactone was required in order to study the origin of the C-13 methyl group (C-18) at the C/D ring junction of cholesterol (**2.19**) in the biosynthetic cyclization of squalene. It was prepared from acetyl chloride by a strategy based on the chemistry of diketene (**2.40**). Treatment of [2-$^{13}$C]acetyl chloride with triethylamine gave the labelled diketene (**2.40**). Reduction with lithium aluminium hydride and acetylation afforded 4-acetoxy[1,3-$^{13}$C$_2$]butan-2-one, which was then converted using unlabelled ketene to [3',4-$^{13}$C$_2$]mevalonolactone (**2.37**). Other [$^{13}$C$_2$]mevalonates have been prepared from [1,2-$^{13}$C$_2$]acetate.

Mevalonolactone labelled at C-3 or on the methyl group has been prepared from C-1- or C-2-labelled ethyl acetate by Grignard reaction with allylmagnesium bromide. Ozonolysis of the double bonds in the Grignard adduct 2.41 and oxidation gave the C$_6$ dicarboxylic acid 3-hydroxy-3-methylglutaric acid. Cyclization of the diacid with acetic anhydride or with *N*,*N*-dicyclohexylcarbodiimide led to the cyclic anhydride, which was then reduced to the 3- or 3'-labelled mevalonolactone **2.37** with sodium borohydride.

Many of these routes have been modified to produce deuteriated or tritiated mevalonates. The preparation of stereospecifically labelled mevalonates is described in Chapter 4. The use of $^3$H:$^{14}$C ratios in the study of terpenoid biosynthesis from mevalonate has shed important light on many biosynthetic steps.


## 2.7 AROMATIC COMPOUNDS LABELLED IN THE RING

Generally labelled [$^{14}$C]benzene, and as a consequence some substituted aromatic compounds, have been prepared by the rearrangement and ring enlargement of [1-$^{14}$C]cyclopentanemethanol followed by dehydrogenation. However, specifically labelled aromatic compounds are often required for metabolic and biosynthetic studies, particularly those which involve the scission of an aromatic ring.

Aromatic substitution reactions often give rise to a mixture of isomers. With labelled compounds this can represent not only a loss of expensive isotope but also a potential source of radioisotopic contamination because of the difficulties of removing traces of isomeric impurities. Consequently, ring syntheses, particularly of highly substituted aromatic compounds, may have some advantages for the synthesis of labelled compounds.

Syntheses involving the Diels–Alder reaction have provided one approach. An example is provided by the synthesis of 4-hydroxy [1,2-$^{13}$C$_2$]benzoic acid (Scheme 2.12). The methyl ester of trimethyl-silyl[2,3-$^{13}$C$_2$]propargylic acid (**2.42**) was prepared from [$^{13}$C$_2$]acetylene. This underwent a Diels–Alder reaction with the diene **2.43** to give the methyl ester of 4-hydroxy[1,2-$^{13}$C$_2$]benzoic acid (**2.44**) and its 2-tri-methylsilyl derivatives, from which the parent 4-hydroxybenzoic acid was obtained.

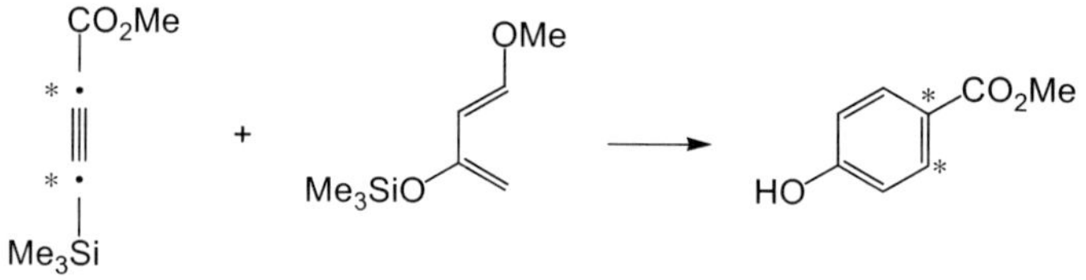

**2.42**              **2.43**                **2.44**

**Scheme 2.12**   Synthesis of 4-hydroxy[1,2-$^{13}$C$_2$]benzoic acid.

The Claisen condensation can be used to create cyclohexanones, and these may be dehydrogenated to phenols, thus providing a route for the specific labelling of these compounds. This is exemplified by the synthesis of 4,6-dihydroxy-2,3-dimethyl[1-$^{14}$C]benzoic acid (**2.47**) (Scheme 2.13), which was required for studies on the biosynthesis of the fungal metabolite mycophenolic acid. The starting materials were 3-methyl-pent-3-en-2-one (**2.45**) and diethyl [2-$^{14}$C]malonate (**2.30**). A Michael addition of the malonate to the unsaturated ketone catalysed by sodium ethoxide followed by a Claisen condensation gave ethyl 2,3-dimethyl-4,6-dioxo[1-$^{14}$C]cyclohexanecarboxylate (**2.46**) in a 'one-pot' reaction. The aromatic ring was then created by dibromination, dehydro-bromination, hydrolysis and hydrogenolysis of the second bromine, to give the required benzoic acid **2.47**.

Another group of ring syntheses involve the addition of a carbanion to an electron-deficient pyrone or pyrylium salt. These are exemplified by the conjugate addition of the anion of diethyl [2-$^{13}$C]malonate to 4$H$-pyran-4-one to give, after hydrolysis and decarboxylation, 4-hydroxy [1-$^{13}$C]benzoic acid. In another synthesis (Scheme 2.14) which afforded 4-nitro[4-$^{13}$C]anisole (**2.49**), [$^{13}$C]nitromethane was added to the pyr-ylium salt **2.48**.

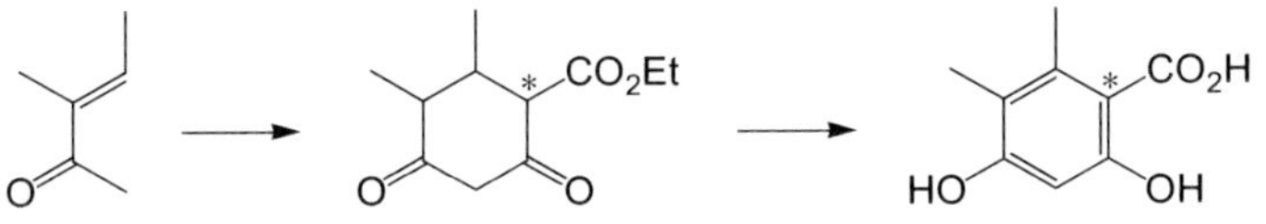

**Scheme 2.13**  Synthesis of 4,6-dihydroxy-2,3-dimethyl[1-$^{14}$C]benzoic acid.

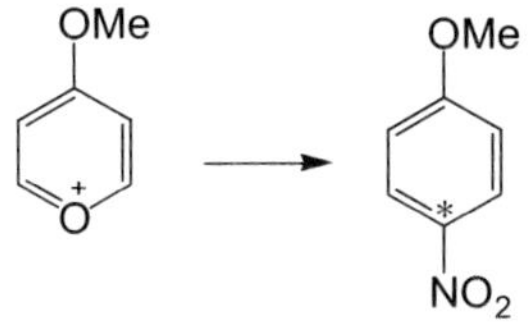

**Scheme 2.14**  Synthesis of 4-nitro[4-$^{13}$C]anisole.

Ring-labelled polycyclic aromatic hydrocarbons have been prepared for metabolic studies and in order to examine their binding to nucleic acids in the context of their carcinogenicity. Many of these syntheses have utilized a Friedel–Crafts cyclization of arylcarboxylic acids. An example is the cyclization of the carboxyl-labelled [$^{14}$C]diphenic acid **2.50** to [9-$^{14}$C]fluorenone (**2.51**) (Scheme 2.15). The labelled acid was prepared from 2-bromodiphenyl by a Grignard reaction utilizing [$^{14}$C]carbon dioxide as the source of the label.

Ring expansion reactions have been used to obtain [9-$^{14}$C]phenanthrene (**2.53**). Carboxylation of the fluorene anion with [$^{14}$C]carbon dioxide and reduction of the fluorene-9-carboxylic acid with lithium aluminium hydride gave 9-fluorene[$^{14}$C]methanol (**2.52**). On treatment with phosphorus pentoxide, this underwent ring expansion to give [9 $^{14}$C]-phenanthrene (**2.53**).

## 2.8   RING-LABELLED HETEROCYCLIC COMPOUNDS

Unlike the syntheses of ring-labelled benzenoid compounds, many of the syntheses of ring-labelled heterocyclic compounds are based on the conventional syntheses of the unlabelled analogues. The syntheses of many heterocyclic compounds containing nitrogen are based either on carbonyl–amine condensation reactions or on nucleophilic substitution reactions. The synthesis of a heterocyclic ring may fall into two stages, first the synthesis of a carbonyl component and second the insertion of the heteroatom in a step which often accompanies the cyclization. These

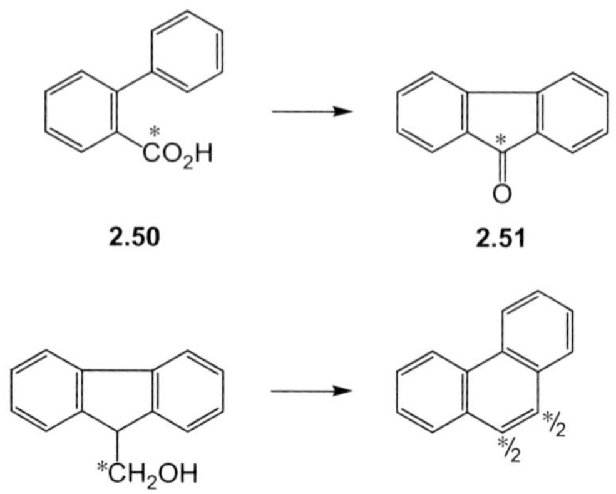

**2.50**          **2.51**

**2.52**          **2.53**

**Scheme 2.15**    Synthesis of carbon-labelled polycyclic aromatic hydrocarbons.

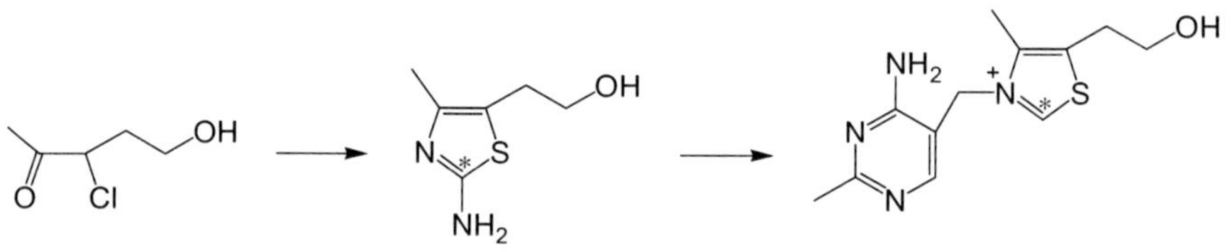

**2.54**          **2.55**          **2.56**

**Scheme 2.16**    Synthesis of $[^{14}C]$thiamin.

stages provide ample opportunity for the regiospecific introduction of a carbon or, as we shall see in Chapter 7, a nitrogen-15 label.

A conventional synthesis which has been used as a labelling procedure is illustrated by the synthesis of $[^{14}C]$thiamin (vitamin $B_1$) (**2.56**) (Scheme 2.16). The condensation of $[^{14}C]$thiourea with 3-chloro-4-oxopentan-1-ol (**2.54**) gave 2-amino-4-methyl-5-(2′-hydroxyethyl)[2-$^{14}C$]thiazole (**2.55**). The 2-amino group was removed by a reduction of the corresponding diazonium compound. The labelled thiazole was then coupled with 2-methyl-4-amino-5-bromomethylpyrimidine to give the vitamin **2.56**.

The metabolism of heterocyclic compounds often involves cleavage of the ring and the separate loss of individual carbon atoms, for example, in the degradation of uric acid. Consequently, it has often been necessary to devise different syntheses of the same molecule in order to introduce a label on different carbon atoms. This is particularly true of the pyrimidine and purine nucleic acid bases.

The syntheses of ring-labelled nucleic acid bases illustrate many of the synthetic methods (Scheme 2.17). Syntheses of pyrimidines and purines

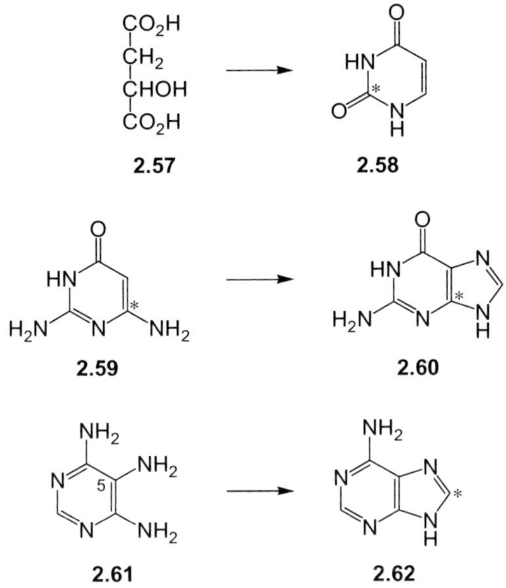

**Scheme 2.17** Synthesis of some [$^{14}$C]nucleic acid bases.

labelled at the C-2 position have made use of carbon-labelled urea, thiourea and guanidine to incorporate the label. These routes are exemplified in the synthesis of [2-$^{14}$C]uracil (**2.58**) by the reaction of [2-$^{14}$C]urea and malic acid (**2.57**) in the presence of sulfuric acid. The malic acid is an *in situ* source of formylacetic acid.

The condensation of a urea with a 1,3-dicarbonyl compound allows labels to be introduced at other centres in the pyrimidine and purine bases. The $C_3$ carbon chain in the ring of a pyrimidine or purine may be labelled by using ethyl cyanoacetate or a malonate derivative. Thus guanidine and ethyl [3-$^{14}$C]cyanoacetate gave 2,4-diamino-6-hydroxy [4-$^{14}$C]pyrimidine (**2.59**). This was subsequently transformed to the labelled guanine **2.60** by nitrosation, reduction and cyclization with formic acid.

The C-8 position of purines can be labelled using [$^{14}$C]formic acid. The synthesis of [8-$^{14}$C]adenine (**2.62**) has been studied fairly thoroughly. 4,6-Diamino-5-nitroso-2-thiopyrimidine was obtained by nitrosation of the condensation product of thiourea and malonitrile. Reduction of this with hydrogen and Raney nickel to remove the sulfur and convert the nitroso group to an amine gave 4,5,6-triaminopyrimidine (**2.61**). The 5-amino group is the more basic of the three amino groups because the other two amino groups are part of an amidine. Hence it was possible to prepare 4,6-diamino-5-[$^{14}$C]formamidopyrimidine by formylation with [$^{14}$C]formic acid. This was cyclized to [8-$^{14}$C]adenine (**2.62**) by heating in

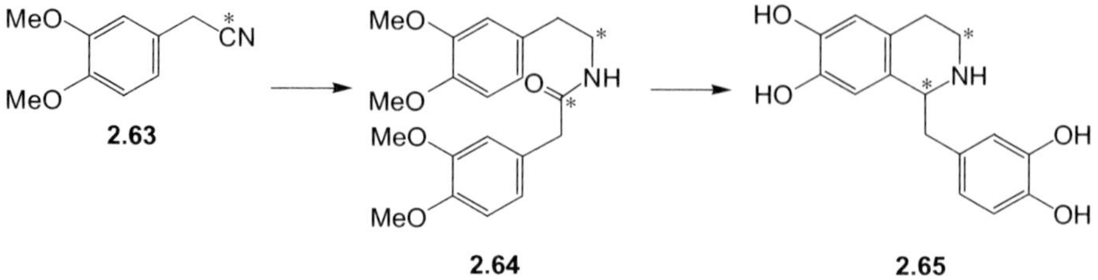

**Scheme 2.18**   Synthesis of labelled benzylisoquinoline alkaloids.

diethanolamine as a solvent. This method has the advantage of introducing the label at the last stage in the synthesis.

Some of the cyclization reactions which lead to isoquinolines involve
electrophilic substitution of the benzenoid ring. The biosynthetic relationship of the benzylisoquinoline alkaloids to the morphine alkaloids
was established using variously labelled benzylisoquinolines such as
[1-$^{14}$C]- and [3-$^{14}$C]norlaudanosoline (**2.65**). These were synthesized by a
general route (Scheme 2.18) from 3,4-dimethoxybenzyl chloride and its
relatives. The label was introduced by nucleophilic substitution of the
benzyl chloride with potassium [$^{14}$C]cyanide. Hydrolysis of the nitrile
**2.63** gave 3′,4′-dimethoxyphenyl[1-$^{14}$C]acetic acid whereas reduction of
this nitrile gave 3′,4′-dimethoxyphenyl[1-$^{14}$C]ethylamine. These provided the means of labelling either of the two halves of the amide **2.64**.
Cyclization of this amide with phosphorus oxychloride gave 3,4-dihy-
dropapaverine. Further reduction and hydrolysis of the methoxyl
groups generated the labelled norlaudanosolines **2.65** required for
the biosynthetic studies. Once this biosynthetic relationship had been
established, further studies on the detailed phenol coupling steps in
morphine biosynthesis required the synthesis of *O*- and *N*-methylben-
zylisoquinoline alkaloids such as reticuline having both tritium and
carbon-14 labels. A general synthetic route to these labelled substrates
was therefore important in facilitating the biosynthetic work.

## 2.9   BIOSYNTHETIC METHODS OF CARBON LABELLING

The specific carbon labelling of the skeleton of complex natural products
by partial degradation and resynthesis can pose considerable difficulties
in terms of devising mild and regiospecific methods. These can sometimes
be overcome by using biosynthetic methods. An example is provided
by the plant hormone gibberellic acid (**2.67**) (Scheme 2.19). The C-17
position of gibberellic acid is potential site for a label. However, the
presence of the adjacent C-13 hydroxyl group and the sensitivity of ring

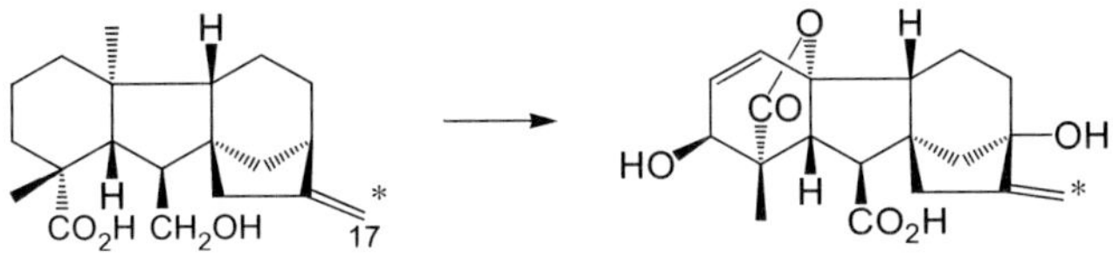

**Scheme 2.19**  Biosynthetic labelling of gibberellic acid.

A of gibberellic acid to both acidic and basic reaction conditions make this a difficult site at which to introduce a carbon label reliably by, for example, a Wittig reaction. Gibberellic acid is produced by the fungus *Gibberella fujikuroi* from less highly hydroxylated precursors. The later stages in the biosynthesis are relatively efficient. An analogue **2.66** of a less highly functionalized key biosynthetic intermediate was more easily labelled chemically at C-17 by a Wittig reaction and then converted into the labelled gibberellic acid **2.67** by the fungus. Labelled gibberellic acid for metabolic studies has also been produced by incubating the fungus with the general isoprenoid precursor [$2$-$^{14}$C]mevalonic acid.

One of the early applications of labelling techniques in biochemistry was in the study of the pathway of carbon in photosynthesis. These studies have formed the basis of methods for the preparation of uni-formly-labelled D-glucose in which a photosynthetic microorganism, *e.g.* a *Chlorella* species, is grown in the presence of [$^{14}$C]carbon dioxide as the sole carbon source. The metabolites were then separated chro-matographically. This technique was sometimes known as 'isotope farming'. The studies by Calvin made it clear that the uniformity of labelling will depend on the time of incubation. Plant leaves, *e.g.* of *Canna* species, have been used for the same purpose. Other compounds, such as myo-inositol and the L-amino acids obtained from protein hydrolysates, have been produced in a labelled form by these methods.

CHAPTER 3

# Labelling with Deuterium and Tritium

## 3.1 INTRODUCTION

The fate of a deuterium or a tritium label in a molecule when it is the substrate for a chemical or enzymatic reaction can provide valuable mechanistic and stereochemical information on the reaction pathway. Furthermore, the effect of isotopic substitution on reaction rates (the kinetic isotope effect) can provide information on the rate-determining step in a multi-stage process.

The replacement of a hydrogen by deuterium modifies the spectroscopic properties of a molecule, particularly the infrared, the mass and the nuclear magnetic resonance spectra. Changes in the IR absorption of a compound arising from the replacement of specific hydrogen atoms by deuterium have played an important role in the assignment of IR absorption bands to particular molecular vibrations. Isotopic substitution can provide a simplification in the $^1$H NMR spectrum by removing various proton signals and couplings. It can also be useful in the identification of various inter- and intramolecular interactions. The location of deuterium and tritium may also be established from the $^2$H and $^3$H NMR spectra. The presence of deuterium also modifies the $^{13}$C NMR spectrum. Carbon–deuterium couplings remain in the proton noise-decoupled spectrum, affecting the intensity of the signals from carbon atoms bearing deuterium. Isotope shifts may also be detected. In biological systems, information can be obtained on protein folding from the identification of resonances in the $^1$H NMR spectrum by the site-specific deuterium labelling of a protein. The effect of deuteriation on the mass

The Organic Chemistry of Isotopic Labelling
By James R. Hanson
© James R. Hanson 2011
Published by the Royal Society of Chemistry, www.rsc.org

spectrum has been particularly useful in elucidating the fragmentation pattern of compounds such as the steroid hormones and in labelling internal standards for the quantitative analysis of components of complex mixtures by chromatographic methods coupled with mass spectrometry.

## 3.2 THE KINETIC ISOTOPE EFFECT

The replacement of an atom by one of its isotopes at, or adjacent to, a reactive centre may have an influence on the rate of reactions involving that centre. This influence is known as the kinetic isotope effect. When a carbon–hydrogen bond is replaced by a carbon–deuterium bond, a reaction in which the carbon–hydrogen bond is broken may have a different rate compared with that of the deuteriated analogue. A condition for isotopic substitution to have a significant effect on the reaction rate is that the relevant C–H bond is being broken in the rate-determining step. This effect is known as the primary kinetic isotope effect. Thus, in the oxidation of cyclohexanol to cyclohexanone by vanadium(v), the reaction of 1-deuteriocyclohexanol [$C^2H(OH)$] proceeded more slowly than the unlabelled material ($k_H/k_D \approx 4$). In the case of the oxidation of benzaldehyde by permanganate, the deuterioaldehyde ($PhC^2HO$) was oxidized at one-seventh of the rate of the benzaldehyde (PhCHO). This led to the conclusions that the alcohol and aldehyde C–H bonds were being broken in the rate-determining steps. In another example, the rates of the bimolecular substitution and elimination reactions of 2-bromo [2-$^2H$]propane and 2-bromo[l,l,1,3,3,3-$^2H_6$]propane in alcoholic sodium ethoxide were compared with those of ordinary 2-bromopropane. Whereas the rates of the substitution reactions were essentially the same, the β-deuterio compound underwent bimolecular elimination 6.7 times more slowly. This is in accordance with a C–$^2H$ bond being broken in the rate-determining step in the elimination reaction. This example is important because it reveals the potential influence that isotope effects may have on the product distribution where there are competing pathways. On the other hand, when a deuteriobenzene was undergoing an aromatic nitration, the rates of the reaction of the protic and deuteriated samples were essentially the same. Taken with other evidence, this led to the suggestion that the formation of the Wheland intermediate may be the rate-determining step and that its subsequent decomposition by the loss of a proton was a relatively fast step and was not rate determining. In a primary isotope effect, the ratio $k_H/k_D$ may be up to 8 and $k_H/k_T$ up to 16, although they are usually considerably less.

Isotopic substitution adjacent to a reactive site may have a much smaller influence on the reaction rate. Where the replacement of hydrogen by deuterium at a bond which is not broken during a reaction brings about a change in the reaction rate, this is known as a secondary kinetic isotope effect. This information can be used, for example, to reveal the extent of hybridization changes in the transition state in elimination reactions. It is not the purpose of this book on synthesis to discuss the mechanistic interpretation of these isotope effects, but nevertheless they do have an influence on the synthesis of labelled compounds in a number of ways.

Clearly, the need to explore the mechanism of a reaction by isotope effects may dictate the site of a label. The presence of an isotope effect may have an unfavourable effect on a labelling reaction. For example, catalytic deuteriation is susceptible to isotope effects. The adsorption of hydrogen *versus* deuterium on the catalyst surface and its transfer to the substrate may favour the lighter isotope. It has often been noted that the incorporation of deuterium from deuterium gas by a catalytic process is less than anticipated. It has also been observed that the incorporation of deuterium on an aromatic ring by the decomposition of an aryl Grignard reagent in 95% deuterium oxide is sometimes less than would be expected.

The primary kinetic isotope effect has been put to synthetic use with deuterium being employed as a 'protecting group'. For example, there have been instances in radical-based syntheses in which a hydrogen atom transfer occurs as an unwanted side reaction quenching a radical and taking place in competition with the required reaction cascade. Replacement of the hydrogen atom involved in this unwanted hydrogen transfer with deuterium significantly reduces the rate of this side reaction and thus suppresses the formation of the unwanted by-product.

Where there are competing biochemical processes and changes in tritium:carbon-14 ratios are being measured between substrates and metabolite, there is also the potential for an isotope effect. This has also, of course, been put to good use in elucidating enzyme mechanisms and some applications will be described later (see, for example, Section 4.6 in Chapter 4).

## 3.3 THE ISOLATION OF DEUTERIUM AND TRITIUM

Deuterium oxide ($^2H_2O$) has a slightly higher boiling point ($101.4\,^\circ$C) than ordinary water and hence it can be obtained by fractional distillation. It will, however, contain some oxygen-18-labelled water. This can be separated by electrolysis. The resulting hydrogen–deuterium mixture is then re-oxidized to water and re-fractionated to give pure deuterium oxide. Deuterium gas can then be obtained from this water by electrolysis or, if only small amounts are required, by reaction with sodium. There is an isotope effect in the electrolysis of water favouring the release of hydrogen and leading to a concentration of deuterium oxide which has also been exploited in the production of heavy water (deuterium oxide).

The enrichment of the deuterium in water can be achieved by a high-pressure exchange between hydrogen and an aqueous suspension of a platinum on carbon catalyst. Hydrogen is passed in a counter-current column through a descending stream of the aqueous suspension of the catalyst. A continuous recycling sequence leads to an enrichment of the deuterium in the water. Another exchange process based on hydrogen sulfide has also been reported.

Tritium is obtained by the neutron bombardment of lithium-6 in the nuclear reaction

$$^{6}Li + n \rightarrow {}^{4}He + {}^{3}H$$

The lithium-6 is held as a lithium–magnesium or lithium–aluminium alloy. Some of the tritium escapes and some is retained as a tritide from which the tritium is released on reaction with acid.

There are very small amounts of tritium which occur naturally in the upper atmosphere, where it is produced by cosmic radiation. Prior to the advent of thermonuclear bombs, it was estimated that there was less than 1 kg of tritium in the world's atmosphere.

Whereas deuterium is a stable isotope, tritium is radioactive, with a half-life of 12.26 years. Tritium is a very weak β-emitter (maximum energy 0.016 meV).

There are four main groups of method that have been used for the introduction of deuterium and tritium into a molecule. These are exchange reactions, protonolysis (deuteriolysis or tritiolysis), reduction and the incorporation of previously labelled fragments (*e.g.* $C^{2}H_{3}-$) into the molecule. Any reaction in which the mechanism involves the protonation of a carbon is potentially a means of introducing deuterium or tritium. In this context, it is also important to consider the stability of a deuterium or tritium label in a compound once it has been introduced. If a reaction is being carried out at another centre in a labelled compound under conditions which can lead to enolization or acid-catalysed exchange, consideration must be given to any possible effect of this on the labelled centre. Tritiated compounds can also undergo slow radiation-induced decomposition.

## 3.4  EXCHANGE REACTIONS IN DEUTERIUM AND TRITIUM LABELLING

### 3.4.1  Heteroatom X–H Exchange Reactions

Deuterium-labelled alcohols and amines of the general formula R–X–$^2$H are widely used in spectroscopic analysis, as reagents for transferring deuterium to a specific site in another molecule and in the elucidation of reaction mechanisms. Simple exchange involving equilibrating $^2$H$_2$O with a solution of an alcohol, acid, amine or amide in an aprotic solvent such as C$^2$HCl$_3$ may be suitable on the NMR scale but it is much less convenient on a preparative scale. Although these exchange reactions are usually rapid, this is not always the case, particularly for strongly hydrogen-bonded systems.

A common preparative procedure which is used for the lower, more volatile aliphatic alcohols is based on the deuterolysis of the metal salt of an alcohol. The sodium or magnesium alkoxides are treated with deuterium oxide and the deuteriated alcohol is then distilled from the residual sodium or magnesium deuteroxide.

The deuterolysis of esters of non-volatile oxy acids by treatment with deuterium oxide provides another group of methods. Borate, carbonate and silicate esters have been employed for this purpose. Carboxylic acid esters and orthoesters have also been used. These reactions are acid catalysed and consequently it may be necessary to use a non-volatile deuteriated inorganic acid such as $^2$H$_2$SO$_4$. Deuteriated acids are best prepared by deuterolysis of an anhydride or acid chloride. If deuterium chloride gas is required as catalyst, it can be prepared by deuterolysis of benzoyl chloride and distilled from the benzoic acid which is also formed.

Although *N*-deuteriated anilines of high isotopic purity can be obtained by a repeated exchange with $^2$H$_2$O, treatment of *N,N*-bistrimethyl-silylaniline with *O*-deuteriomethanol gives *N,N*-dideuterioaniline in high isotopic purity in one step.

### 3.4.2  General C–H Exchange Reactions

The primary sources for many tritium labels are tritium gas or tritiated water. There are a number of procedures for the general exchange of hydrogen atoms for tritium atoms. In the Wilzbach procedure, a substance is exposed to tritium gas for a period of days or even weeks and a slow C–H exchange takes place. The exchange may be catalysed by charcoal or a platinum or palladium catalyst. Although the process is simple and a significant number of compounds have been tritiated in this way, the procedure suffers from a number of limitations arising mainly from the lack of stability of compounds to the exchange conditions. For example, double bonds may be reduced and halogens, particularly bromine and iodine, may be hydrogenolysed. The decomposition may lead to the introduction of a significant amount of label into the decomposition product. There can then be the problem of removing highly labelled trace impurities in order to achieve radiochemical purity. However, with microwave radiation to facilitate the exchange and modern HPLC purification techniques combined with the facilities for handling tritium gas, this can be a useful technique.

Another method of generally labelling a compound utilizes a Lewis acid catalyst and tritiated water. This procedure is particularly useful for labelling aromatic compounds. However, tritiated water of the highest isotopic abundance is seldom used because of the danger of radiation damage to the compounds concerned. For example, styrene will undergo radiation-induced polymerization under these conditions. Alkanes which contain a tertiary hydrogen, such as 2,3-dimethylbutane, undergo exchange in the presence of ethylaluminium dichloride and tritiated water. $^3$H NMR studies have shown that the label is spread throughout the molecule. The procedure can also be used to introduce deuterium into compounds such as aromatic hydrocarbons.

### 3.4.3  Site-specific C–H Exchange Reactions

Although a few C–H exchange reactions, *e.g.* that of H-3 in 4-hydroxy-coumarin, have been reported to take place without catalysis, the majority of site-specific C–H exchange reactions require either acidic or basic catalysis. The protons of an aromatic ring will undergo exchange with deuterium or tritium from $^2$H$_2$O or $^3$H$_2$O in the presence of an acid catalyst at sites and rates which are dependent on the activating or deactivating effect of other substituents attached to the ring. In one of the earliest deuteriation experiments, reported in 1934, hexadeuteriobenzene was prepared by heating benzene with deuteriosulfuric acid. In 1936, Ingold showed that the *ortho* and *para* hydrogens of phenol underwent exchange with deuterium when the phenol was heated with alkaline deuterium oxide. Studies with *p*-cresol showed that exchange adjacent to the phenol could be carried out under milder conditions by heating with triethylamine as the base. Indeed, a simple method of labelling some complex phenols such as morphine involved just heating the phenol in dimethylformamide or dimethyl sulfoxide with deuterium oxide or

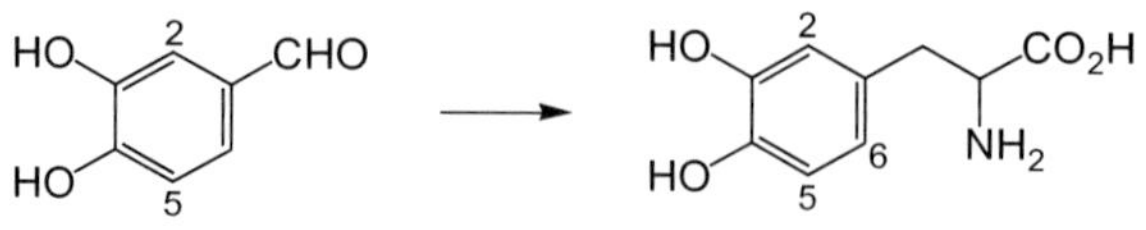

**3.1**                                                   **3.2**

**Scheme 3.1**   The labelling of 3,4-dihydroxyphenylalanine.

tritiated water. In a simple sequence of reactions (Scheme 3.1), the amino acids tyrosine and 3,4-dihydroxyphenylalanine (**3.2**) were labelled *ortho* to the phenolic hydroxyl groups by deuterium and tritium. 4-Hydroxybenzaldehyde or protocatechuic aldehyde (**3.1**) were heated in deuterium oxide or tritiated water in the presence of base to introduce the label on to the aromatic rings. The amino acids, for example 3,4-dihydroxy[2,5-$^3$H$_2$]phenylalanine (**3.2**), were then synthesized by condensation of the aldehyde with hydantoin, reduction and hydrolysis.

The exchange can also take place under acid catalysis. Thus DL-3,4-dihydroxy-[2,5,6-$^3$H$_3$]phenylalanine (**3.2**) was obtained by heating the DL-acid with tritiated hydrochloric acid at 100 °C for 3 h. The phenolic steroidal hormone estradiol afforded [2,4-$^2$H$_2$]estradiol (**3.3**) on heating with [O-$^2$H]methanol and 10% deuteriosulfuric acid.

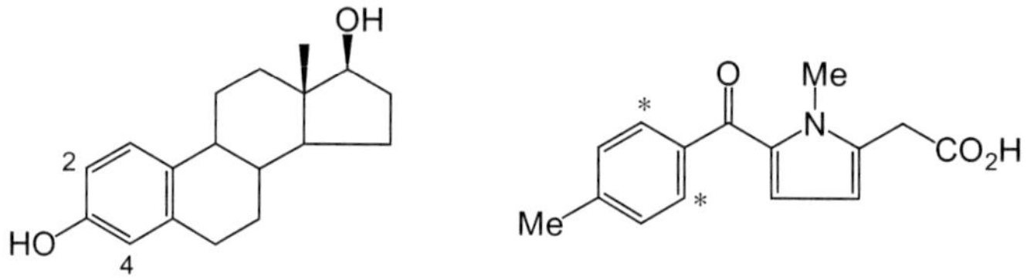

**3.3**                                                   **3.4**

The directing and deactivating effect of aromatic nitro substituents is revealed by the deuteriation of 4-nitrotoluene, which required heating for 24 h at 90 °C in deuteriosulfuric acid to give the 2,6-dideuterio compound. However, it is worth remembering that a nitro group can be transformed *via* the corresponding amine and diazotization into a variety of other substituents. Studies on the kinetics of the reverse reaction, acid-catalysed detritiation, have shed considerable light on aromatic substituent effects, enabling partial rate factors to be measured for particular sites.

Catalytic *ortho*-directed hydrogen isotope exchange between isotopically labelled water and substituted aromatic compounds has become a very useful procedure. The rate of these exchange reactions has been enhanced by the use of microwave irradiation. Heating a carb-oxylic acid or a salt such as sodium benzoate with a mixture of deu-terium oxide and *N,N*-dimethylformamide in the presence of a rhodium trichloride catalyst led to the regioselective deuteriation of the *ortho* position. The reaction has been extended to the use of tritium oxide and to deuterium or tritium gas. The more recent introduction of novel catalysts such as cyclooctadienyl (cod) iridium(I) penta-1,3-dionate has extended the scope of the reaction. Thus 4-phenylbenzoic acid gave 4-phenyl-[2,6-$^2$H$_2$]benzoic acid with 97% deuteriation at the two pos-itions adjacent to the carboxylic acid. A series of iridium catalysts of the general type [Ir(PR$_3$)$_2$(cod)]$^+$BF$_4^-$ have been examined and found to be very effective in mediating exchange *ortho* to the carbonyl group of a substituted acetophenone. An example of their use is in the labelling of the non-steroidal anti-inflammatory drug (NSAID) tolmetin (**3.4**) with tritium.

A C–H on a heteroaromatic ring can undergo exchange under a variety of conditions. In general, the exchange reactions of π-excessive heteroaromatic rings such as pyrroles are easier than those of π-deficient systems such as pyridines. Although they undergo rapid exchange, pyrroles tend to undergo decomposition in the presence of acid. Base-catalysed exchange which takes place at both the α- and β-positions is the preferred method of labelling. Pyrazoles have been labelled by base-catalysed isotopic exchange in superheated deuterium oxide under pressure. In the presence of a mild base (K$_2$CO$_3$), 2,4-dimethylpyrazole gave the [1,3]dideuterio product whereas the fully deuteriated product was obtained in the presence of a stronger base.

Base-catalysed exchange reactions have been used to prepare deu-teriated furans. When furan-2-carboxylic acid was heated with deuter-ium oxide containing 40% sodium deuteroxide for 16 h, exchange occurred at both the α- and β-positions. Tetradeuteriofuran was pre-pared from this product by decarboxylation of the sodium salt of the acid with mercuric chloride to form the 2-mercurichloride. The orga-nomercurial derivative was then cleaved with deuterium chloride.

The C-2 proton of imidazole and benzimidazole (**3.5**) undergoes ex-change in deuterium oxide on heating even without catalysis. When the imidazole ring forms part of the purine ring system **3.6**, this proton (H-8) exchanges fairly rapidly. The base-catalysed exchange of the protons at C-2 and C-5 of thiazoles with deuterium has been studied thoroughly in the context of the role of the thiazole ring in vitamin B$_1$ (thiamin) (**3.7**). The presence of the sulfur facilitates the formation of a carbanion at C-2. This carbanion on the thiazole ring is involved in the reactions of the vitamin as a coenzyme.

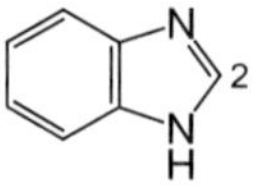

**3.5** 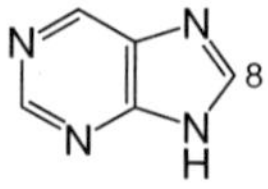              **3.6**

Because of their lower reactivity, the site-specific deuteriation of pyridines requires various indirect methods such as reductive debromination and decarboxylation. Pyrimidines such as uracil (**3.8**) undergo exchange at C-5.

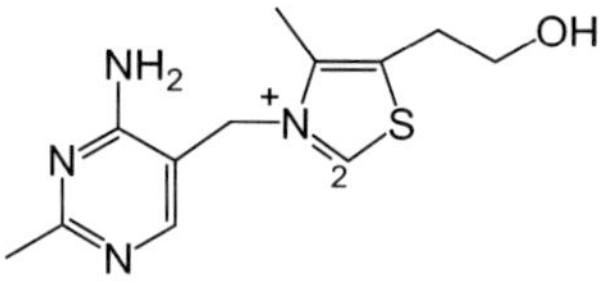

**3.7** 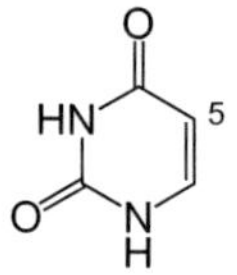              **3.8**

A useful exchange reaction of a methylene adjacent to an electron-deficient nitrogen leads to the preparation of deuteriated diazomethane (**3.10**). When the preparation (Scheme 3.2) is carried out by heating the common diazomethane precursor N-methyl-N-nitrosotoluene-p-sulfonamide (Diazald®) (**3.9**) with sodium deuteroxide in deuterium oxide and deuteriated carbitol, the product which is distilled out of the solution is [$^2$H$_2$]-diazomethane (**3.10**).

The complexity of a number of acid-catalysed reactions and the extent to which various centres are involved have been studied by carrying out the reactions in a deuteriated medium. Useful mechanistic information concerning the dienol–benzene and dienone–phenol rearrangements has been obtained this way. It has also provided a route for the preparation of some specifically deuteriated compounds. Measuring the rates of incorporation or loss of tritium in aromatic substitution has played an important role in the study of the mechanism of aromatic substitution. Although the results of these mechanistic studies are outside the scope of this book, it is worth noting that some of these reactions have been used in a preparative context.

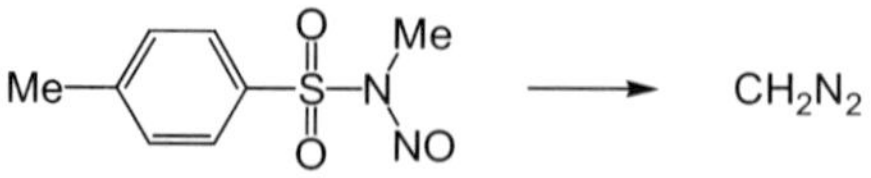

**3.9**                                   **3.10**

**Scheme 3.2**   The synthesis of [$^2$H$_2$]diazomethane.

A procedure which has been used to introduce deuterium in a site-specific manner on to an aromatic ring is to use the directing effect of an aromatic amine in an acid-catalysed deuteriation. The amino group is then removed by diazotization and reduction. Alternatively, the site of the amino group can be labelled with deuterium by carrying out a diazotization and reducing the diazonium group with deuteriohypophosphorous acid ($^2H_3PO_2$). However, there is evidence to suggest that these reactions are not as site specific as a simple interpretation of the reaction would imply. The replacement of a substituent by a deuterium is not restricted to the diazonium salt. Other reductive methods are discussed later.

### 3.4.4 Exchange Reactions of Carbonyl Compounds

The synthesis of α-deuteriated and tritiated carbonyl compounds by exchange reactions is a widely used technique. These reactions take place with either acid or base catalysis although the latter is the more widely used. Some reactions are fairly fast and complete within 30 min to 1 h in refluxing deuterium oxide or [$O$-$^2H$]methanol containing sodium deuteroxide whereas others may take some days. These differences arise from the effect of structure on the rate and direction of enolization of a ketone.

[$^2H_6$]Acetone has been prepared from acetone by a series of repetitive exchange reactions with deuterium oxide in the presence of anhydrous potassium carbonate. In one procedure each exchange reaction was carried out at room temperature for 36 h. After each exchange reaction, the acetone-$d_x$ was fractionally distilled from the deuterium oxide and used as the starting material for the next reaction.

The ease of deuterium exchange reactions adjacent to carbonyl groups is illustrated in Scheme 3.3 by the preparation of (3$RS$)-[$^2H_5$]linalool (3.12), which was required for a study of the biosynthesis of the lilac aldehydes and alcohols produced by *Syringia vulgaris*. [1,1,1,3,3-$^2H_5$]-6-Methylhept-5-en-2-one (3.11) was prepared from the unlabelled ketone by exchange using [$O$-$^2H$]methanol containing sodium methoxide. It was

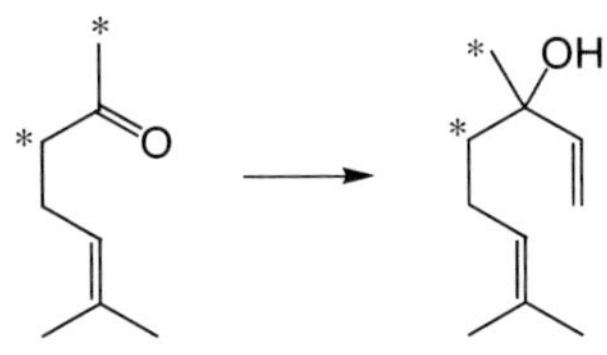

**3.11**        **3.12**

**Scheme 3.3**   The synthesis of [$^2H_5$]linalool.

then converted to the labelled linalool **3.12** by a Grignard reaction with vinylmagnesium bromide. In this example, a simple dissection of the target molecule revealed a ketone which could be used as an activating group for the introduction of the label. This strategy for introducing a label into the core of a molecule has been used in a number of situations.

The exchange reactions of cyclic ketones have been thoroughly examined both in the context of studies into their enolization and in the identification of ions in their mass spectra. In the steroid series **3.13**, the C-2 and C-4 methylenes adjacent to a C-3 ketone undergo exchange fairly rapidly but a difference is shown by a C-7 and a C-11 ketone. In the presence of a C-7 ketone, the H-6 methylene exchanges fairly rapidly with deuterium, but H-8 takes much longer, reflecting the favoured direction of enolization. With a C-11 ketone, the axial H-9 and H-12α protons are exchanged much more rapidly than the equatorial H-12β proton, reflecting the stereochemistry of protonation (deuteriation) of the enol.

When a cyclopentanone is part of a bridged ring system as in camphor (**3.14**), exchange of the adjacent *exo*-proton takes place much more rapidly under base-catalysed conditions than does exchange of the *endo*-proton. It is possible to obtain both deuteriated epimers since the difference in reactivity extends to the reverse dedeuteriation reaction of the dideuteriated species.

**3.13**                          **3.14**                          **3.15**

αβ-Unsaturated ketones and esters undergo exchange at the γ-position in addition to the α-positions. Deuteriation at the α-position takes place by an addition/elimination mechanism whereas exchange at the γ-position involves enolization and re-ketonization, Thus ethyl cinnamate, on treatment with [$O$-$^2$H]ethanol containing a small amount of sodium ethoxide gave the α-deuterio compound. The unsaturated ketone testosterone (**3.15**) is labelled at C-2, C-4 and C-6 on treatment with [$O$-$^2$H]methanol containing sodium deuteroxide. Some stereochemical preferences are revealed when the deuteriation is carried out under acid-catalysed conditions. Thus deuteriation of the αβ-unsaturated ketone and the corresponding enol ether gave the axial 6β-deuterio compound.

The C-2 methylene of a β-keto ester such as ethyl acetoacetate undergoes a facile exchange in deuterium oxide. This has been combined with a reduction of the β-carbonyl group by adding a yeast (*Saccharomyces cerevisiae*) to the deuterium oxide to give ethyl (*S*)-3-hydroxy [2-$^2H_2$]butyrate. There was enough NADPH in the yeast to ensure that very little label was introduced at C-3 during the reduction. The labelled butyrate was in turn converted to the *N*-acetylcysteamine thioester for use in biosynthetic studies.

The temporary conversion of a centre to an exchangeable position is a useful labelling procedure. For example, the C–H of a formyl group (CHO) is not normally an exchangeable position, but this outcome can be achieved by converting it to a suitable derivative. Thus the 2-H of a 1,3-dithiane derived from an aldehyde, *e.g.* benzaldehyde, is sufficiently acidic to allow the formation of a C-2 carbanion which is stabilized by the sulfur atoms. The carbanion can be quenched with deuterium oxide and the labelled aldehyde regenerated with mercuric chloride. Another derivative which was used in the preparation of [$^2H$]benzaldehyde involved the condensation of the aldehyde with morpholine to give an iminium salt. Addition of the cyanide ion led to a morpholinonitrile (**3.16**), in which the original formyl C–H is now rendered acidic by the nitrile. The carbanion may be created with sodium hydride and quenched with deuterium oxide. Hydrolysis of the derivative regenerated the labelled aldehyde. Other methods of labelling the formyl group of an arylaldehyde involve decarboxylation of a benzoylformic acid or an α-imino acid or reduction of an iminium salt (see Sections 3.5 and 3.6.6).

## 3.4.5 Exchange Reactions in the Preparation of Some Deuteriated Solvents

Deuteriated solvents are widely used in NMR spectroscopy. Deuteriochloroform may be prepared by the decomposition of chloral hydrate in sodium deuteroxide or by the simple exchange reaction of chloroform using a solution of sodium deuteroxide in deuterium oxide in the presence of a phase transfer catalyst. Dideuteriomethylene chloride is prepared by a similar exchange. Hexadeuterioacetone and hexadeuteriodimethyl sulfoxide are also prepared by exchange reactions using deuterium oxide.

**3.16**                    **3.17**

## 3.5 DECARBOXYLATION REACTIONS IN DEUTERIUM AND TRITIUM LABELLING

Decarboxylation reactions are often accompanied by protonation of the site of the former carboxyl group and therefore they provide a potential method for deuteriation and tritiation. Malonic acids readily undergo decarboxylation and this has been used in the context of preparing labelled compounds. [$^2$H$_4$]Acetic acid has been prepared by the thermal decarboxylation of [$^2$H$_4$]malonic acid. The deuteriated malonic acid was prepared by the addition of deuterium oxide to carbon suboxide ($C_3O_2$). The latter was prepared by the pyrolysis of diacetyltartaric acid. [2,2-$^2$H$_2$]Succinic acid, which was required for biosynthetic studies, was prepared by heating triethyl 1,1,2-ethanetricarboxylate in a sealed tube at 160 °C.

The thermal decarboxylation of the calcium salts of benzenecarboxylic acids in the presence of calcium deuteroxide has been used to prepare variously deuteriated aromatic compounds. The method has also been used to prepare 2-, 3- and 4-deuteriated pyridines. The decarboxylation of aromatic acids is copper catalysed. The effectiveness of this as a deuteriation procedure has been enhanced by the presence of quinoline and by the use of microwave irradiation. An example of its application has been in the labelling of an aromatic aldehyde by decarboxylation of the corresponding benzoylformic acid (**3.17**).

The decarboxylation of aromatic α-imino acids (**3.18**) catalysed by tributylphosphine in the presence of acetic acid as a proton donor gave the corresponding aromatic imine **3.19** (Scheme 3.4). Use of deuteriated or tritiated acetic acid gave the labelled imines from which the aromatic aldehyde labelled in the formyl group could be obtained by hydrolysis.

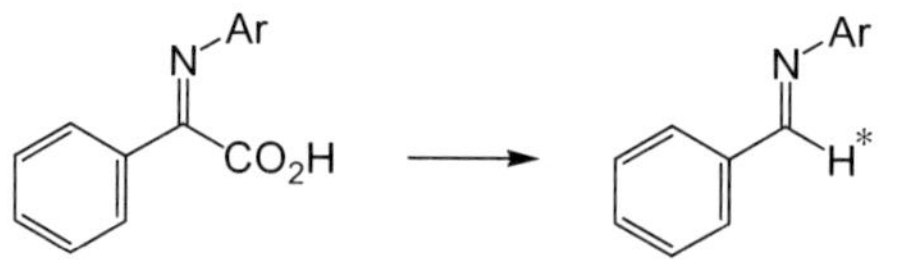

**3.18**                              **3.19**

**Scheme 3.4**   The synthesis of labelled aldimines.

## 3.6 REDUCTION REACTIONS

### 3.6.1 Reducing Agents

The reduction of a functional group such as a carbonyl group, an epoxide, an alkyl halide or an alkane in order to introduce a deuterium or tritium atom is part of many labelling sequences. Reducing agents fall into four major classes based on their mechanism of action. Each class includes methods for introducing a deuterium or a tritium label. The hydride reagents such as sodium borodeuteride, sodium borotritide and lithium aluminium deuteride introduce deuterium or tritium as a nucleophile. Highly deuteriated and tritiated samples of lithium, sodium and potassium borohydride have been prepared by exchange reactions with deuterium or tritium gas at 200–350 °C. Although sodium borohydride does not undergo exchange with deuterium oxide, lithium borohydride does react. In the 'dissolving metal' family such as sodium dissolving in $[O\text{-}^2H]$ethanol, electrons are first delivered from the metal to the substrate and the reaction is then completed by the addition of deuterium to a radical or anionic species. The catalytic group of reactions involve the delivery of deuterium or tritium from the catalyst to an adsorbed substrate. The final group of reagents are those that transfer deuterium or tritium from a labelled reagent to the substrate, often by a cyclic mechanism. This family includes a number of enzymatic methods where the source of the label is labelled NADH or NADPH.

Within each family of reagents there are a graded series of reactivities. For example, various hydrides can be used to differentiate between carbonyl groups. This is based on the differing sensitivity of the carbonyl groups to nucleophilic attack, on changes in the solvent and on the presence of additional metal salts which may chelate to the carbonyl oxygen.

The stereochemical outcome of these reactions is a reflection of their mechanism of action. To a first approximation, the stereochemistry of the reduction by hydride reagents is dominated by the trajectory of the approach of the hydride to the carbonyl group. On the other hand, the stereochemistry of reduction by a dissolving metal reagent is determined by the freedom of the carbanion derived from the carbonyl group by the addition of the electrons to take up the most stable conformation prior to protonation (deuteriation or tritiation). The stereochemistry of reduction by catalytic reagents is determined by the relative ease of access of the different faces of the alkene or carbonyl group to the catalyst. The stereochemistry of the final group of reagents is determined by the steric interactions in the complex that is formed as the hydride transfer takes place. Although the reduction of a ketone may generate a particular labelled isomer of an alcohol, it is worth remembering that its labelled epimer may be generated by the nucleophilic substitution of a suitable derivative such as the methanesulfonate, triflate or toluene-*p*-sulfonate by a hydroxide or acetate anion.

### 3.6.2   The Reduction of Carbonyl Groups

The reduction of an aldehyde or a ketone by sodium borodeuteride or sodium borotritide is a convenient way of labelling an alcohol. Thus the terpenoid precursors isopentenyl diphosphate, geranyl diphosphate and the $C_{15}$ analogue farnesyl diphosphate have each been labelled at C-1 by reducing the corresponding aldehyde with sodium borodeuteride or tritide followed by a phosphorylation. Dideuteriation has been carried out by reducing an ester of the corresponding acid with lithium aluminium deuteride.

### 3.6.3   The Hydrogenolysis of Aliphatic Substituents

The nucleophilic substitution of a methane- or toluene-*p*-sulfonate by a hydride from sodium borohydride or lithium aluminium hydride is a useful way of stereospecifically introducing a label. Typical of a bimolecular nucleophilic substitution, the reaction proceeds with inversion of configuration. The label can either be introduced by the reduction of a carbonyl compound to form the labelled alcohol precursor of the sulfonate or it can be introduced by the nucleophilic substitution of the sulfonate, providing an opportunity to generate epimeric labelled centres.

   Labelled dihydrosqualene (**3.20**) was required for steroid biosynthetic studies. A trisnorsqualene aldehyde was prepared by selective epoxidation of squalene *via* a bromohydrin, hydrolysis of the epoxide to a diol and cleavage of the diol with sodium periodate to form the aldehyde. The three carbons of the isopropyl group were restored by a Grignard reaction with isopropylmagnesium bromide. The hydroxyl group at C-3 was converted to its toluene-*p*-sulfonate and hydrogenolysed with lithium aluminium tritide to generate the labelled dihydrosqualene (**3.20**). In this example, it is worth noting that the label was introduced in the last step in the synthesis.

### 3.6.4   The Catalytic Reduction of Alkenes and Alkynes

One of many examples of catalytic deuteration is the preparation of [$^2H_8$]chlorhexidine (**3.21**). This topical antiseptic has widespread use in Hibitane® preparations and along with cetrimide in Savlon® creams. The hydroxyl groups of hexa-2,4-diyne-1,6-diol were protected as their THP ethers and then the alkynes were reduced using deuterium gas and a tris(triphenylphosphine)rhodium(I) chloride catalyst. The protecting groups were removed and the resultant [2,3,4,5-$^2H_8$]hexane-1,6-diol was converted to the 1,6-diamine and thence to the biscyanoguanidine with sodium dicyanamide. Finally, the *p*-chloroaniline units were added to give the [$^2$H$_8$]chlorhexidine (**3.21**).

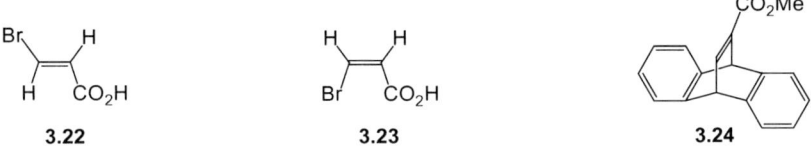

**3.20**                    **3.21**

The reduction of alkenes by catalytic hydrogen transfer from a donor such as labelled formate in the presence of Wilkinson's catalyst has been used to introduce a label. The rate of this hydrogenation reaction has been enhanced by microwave irradiation. Various donors can give rise to different labelling patterns in the reduction of unsymmetrical alkenes. For example, dihydrocinnamic acid has been obtained from cinnamic acid with the deuterium on either the α- or β-carbon atoms depending on whether potassium [$^2$H]formate or deuterium oxide was the source of the label.

Stereospecifically deuteriated or tritiated acrylic acid has made a useful building block for the synthesis of more complex compounds of biological importance. The partial catalytic hydrogenation of alkynes over a Lindlar catalyst (Pd/CaCO$_3$) gives a *cis*-alkene. Although the partial catalytic deuteriation of the acetylenic acid propiolic acid gave predominantly the *cis* product, the reduction was not completely stereospecific. Hence various methods were developed to overcome the problem (Scheme 3.5). The (*E*)- and (*Z*)-3-bromopropionic acids **3.22** and **3.23** have been prepared by heating propiolic acid with hydrogen bromide. By adjusting the reaction conditions, the formation of either the *E* or the *Z* geometric isomers can be selectively favoured. The (*E*)-and (*Z*)-3-bromopropionic acids can be separated and obtained pure. Reduction of the bromo compounds with sodium in deuterium oxide or tritiated water proceeds stereospecifically to give the (*E*)- or

**3.22**                **3.23**                **3.24**

**Scheme 3.5**   The synthesis of labelled acrylic acids.

(*Z*)-[3-$^2$H$_1$ or $^3$H$_1$]acrylic acids with retention of configuration. In another approach, the Diels–Alder adduct **3.24** of the methyl ester of propiolic acid and anthracene was hydrogenated (deuteriated) over a palladium catalyst in a stereospecifically cis manner. Pyrolysis then brought about a retro-Diels–Alder reaction and liberated the *cis*-[2,3-$^2$H$_2$]acrylic acid. The anthracene not only protected the system against over-reduction but also enhanced the *cis* addition. Mono-deuteriopropiolic acid could be used in this sequence with a catalytic hydrogenation to give (*E*)-[3-$^2$H]acrylic acid. The mono-deuteriopropiolic acid was prepared by the decarboxylation of the monopotassium salt of acetylenedicarboxylic acid by heating in deuterium oxide at 100 °C. The labelled acrylic acids have been used to prepare labelled amino acids and shikimic acid.

### 3.6.5   The Hydrogenolysis of Aryl Halides

The hydrogenolysis of aryl halides is an important method for specifically deuteriating and tritiating an aromatic ring. The substitution can be carried out with metal hydrides in the presence of a catalyst using sodium [$^3$H]borohydride or lithium aluminium deuteride or catalytically over palladium on charcoal using deuterium or tritium gas. Thus tritiolysis of 4-bromobiphenyl occurred with sodium [$^3$H]borohydride in the presence of tetrakis(triphenylphosphine)palladium(0). This procedure was used to label lysergic acid diethylamide (LSD) (**3.26**) by displacement of an indole 2-bromo substituent (**3.25**).

| 3.25 | R = Br |
| 3.26 | R = $^3$H |

| 3.27 | R = Br |
| 3.28 | R = $^3$H |

During studies on the labelling of carcinogenic aromatic hydrocarbons, it was shown that when lithium aluminium deuteride was used for deuterolysis, the overall deuterium incorporation was dependent on the conditions. In particular, complete deuterium incorporation was only achieved when deuterium oxide was also used in the work-up. Attention to the detail of the reaction conditions was also revealed by studies on the tritiation of fluoxetine (Prozac®). When [³H]fluoxetine (**3.28**) was prepared by catalytic reduction of the bromo compound **3.27** with tritium gas in the presence of palladium on charcoal in ethanol, the reaction was accompanied by a significant amount of hydrogenolysis of the benzylic carbon–oxygen bond. This side reaction was suppressed by carrying out the reduction in pyridine.

Normally, the tritiodehalogenation of aryl bromides and iodides takes place more readily than that of chlorides. This is exemplified by the preparation of the anti-hypertensive agent clonidine (**3.32**) labelled with tritium (Scheme 3.6). Bromination of *p*-aminoclonidine hydrochloride (**3.29**) gave the dibromo compound **3.30**. The amino group was then replaced with iodine by diazotization and treatment with potassium iodide to give 3,5-dibromo-4-iodoclonidine (**3.31**). Catalytic tritium dehalogenation over 10% palladium on charcoal gave [*phenyl*-³H]clonidine hydrochloride (**3.32**) in which displacement of the bromine and iodine, but not the chlorine atoms, had occurred.

Although reductive dehalogenation of aryl bromides and iodides proceeds more readily that that of the chlorides, the availability of starting materials may dictate that the latter are used. The chemical yield may be low but, provided that a good separation can be achieved, material of high specific activity can be obtained. For example, a sample of (*S*)-methamphetamine with a high specific activity was prepared starting from 2,6-dichlorobenzaldehyde. This was converted to racemic 2,6-dichloroamphetamine (**3.33**) by condensation with nitroethane and reduction. *N*-Methylation was carried out by a reduction of the carbamate. The racemic 2,6-dichloromethamphetamine was resolved with

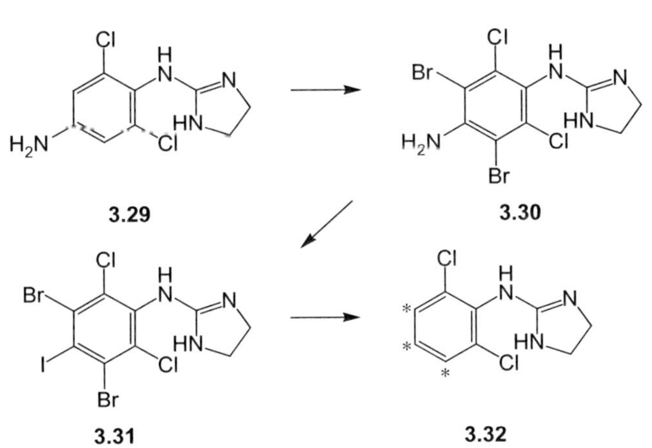

**Scheme 3.6**   The synthesis of [*phenyl*-³H]clonidine.

(+)-dibenzoyltartaric acid. Although the chemical yield was low, (S)-[³H]methamphetamine of high specific activity was obtained by tritiolysis of the chlorine atoms over 20% palladium on charcoal. The difference in the reactivity of aryl chlorides and bromides was exploited in a preparation of (S)-[³H]amphetamine. The bromine of 3,5-dichlorobromobenzene was displaced by treatment with butyllithium and (R)-propylene oxide to give the chiral dichlorophenylpropan-2-ol (**3.34**). The hydroxyl group was then converted *via* the toluene-*p*-sulfonate and the azide to the amine **3.35**. Reductive dechlorination with tritium gas over palladium on charcoal gave the (S)-[³H]amphetamine (**3.36**).

| | | |
|---|---|---|
| **3.33** | **3.34** | **3.35**   R = Cl |
| | | **3.36**   R = ³H |

### 3.6.6   The Reduction of Iminium Salts

The reduction of iminium salts has provided a means of introducing labels into organic compounds. A method for converting a carboxylic acid such as benzoic acid into the corresponding deuteriated aldehyde has been devised based on the reduction of 1,3-dimethylbenzimidazolium salts (**3.37**) with sodium borodeuteride. The salts were prepared by reaction of the carboxylic acid with *o*-phenylenediamine and quaternization with methyl iodide. Reduction and hydrolysis gave the labelled aldehyde. A similar procedure is based on the reduction of dihydro-1,3-oxazines (**3.38** and **3.39**).

| | | |
|---|---|---|
| **3.37** | **3.38** | **3.39** |

### 3.6.7   The Hydrogenolysis of Organometallic Compounds

The mechanism of some of the methods described in the previous sections involve the formation of organometallic intermediates. However, there are a number of methods which involve the discrete formation of an organometallic compound, typically a Grignard reagent or an organolithium compound, in the first step followed by its decomposition with deuterium oxide or tritiated water:

$$RBr + Mg \rightarrow RMgBr \rightarrow RH^*$$

Although these methods have been used with alkyl halides, they have found more widespread application with aryl and vinyl halides. The location of the organometallic moiety is defined by the initial site of the halogen, typically bromine. For example, [2-$^2$H]benzyl chloride was required in a study of the formation of the tropylium ion. The deuterium was introduced on to the aromatic ring by decomposition of the Grignard reagent from 2-bromotoluene with deuterium oxide. This was then followed by a selective chlorination of the methyl group of the [2-$^2$H]toluene with sulfuryl chloride. A large number of specifically tritiated aromatic compounds were required for a study of substituent effects on hydrogen exchange reactions in which the partial rate factors for acid-catalysed detritiation were measured. In these studies, it was important that the exact location of the label was known, something which could be achieved by starting with a bromo compound of known orientation. The tritium labels were introduced by decomposition of the Grignard or organolithium compound with [$^3$H]water. For example, [4-$^3$H]toluene was prepared from 4-bromotoluene.

The reduction of aryl bromo compounds with a Raney copper–aluminium alloy in 10% sodium deuteroxide in deuterium oxide is another regiospecific way of introducing deuterium. This procedure has been used with bromophenols and bromobenzoic acids. All 19 possible mono- and polydeuteriated benzoic acids have been prepared for a $^1$H NMR study by this method.

### 3.6.8 Hydroboration

The addition of borane to an alkene proceeds in a *cis* manner to generate an alkylborane. The electron-deficient borane behaves as a Lewis acid and the addition to a trisubstituted alkene proceeds with a regiochemistry in which the borane becomes attached to the less highly substituted carbon. The value of these alkylboranes in synthesis lies in their reactions with various reagents. The most useful reaction is with alkaline hydrogen peroxide. The hydroperoxide anion is a powerful nucleophile which adds to the electron-deficient borane to form an 'ate' complex. Rearrangement and decomposition of this leads to an alcohol in which the hydroxyl group has replaced the boron of the borane. The overall consequence of this reaction is a *cis*-Markovnikov hydration of an alkene. The regiochemistry of the hydration should be contrasted with the acid-catalysed hydration of an alkene and with the reduction of epoxides.

The deuteriated borane–tetrahydrofuran complex has been prepared by treating a suspension of sodium borodeuteride in glyme with boron trifluoride etherate and passing the resultant diborane into tetrahydrofuran. The tetrahydrofuran solution has been stabilized with amines such as *N*-isopropyl-*N*-methyl-*tert*-butylamine. Sodium borotritide can replace the sodium borodeuteride.

An example of the use of tritiohydroboration comes from the labelling at C-6 of the *ent*-7α-hydroxykaurenoic acid (**3.40**) precursor of the gibberellin plant hormones in a study of the ring contraction step in their biosynthesis. Hydroboration using tritiated borane and oxidation of the borane from 16-ketal of methyl 16-oxo-17-norkaur-6-en-19-oate led to the stereo- and regiospecific introduction of the required label. Hydrolysis of the protecting groups and introduction of the 16(17)-alkene by a Wittig reaction then afforded the labelled biosynthetic intermediate.

The protonolysis (deuterolysis) of boranes with deuterioacetic acid provides a means of introducing deuterium into the steroids. Some further stereochemical aspects of the reactions of boranes in labelling compounds are discussed in Chapter 4.

**3.40**                                          **3.41**

## 3.7  DEUTERIUM- AND TRITIUM-LABELLED ALKYL HALIDES

Alkyl halides play an important role in the introduction of both carbon and hydrogen labels into organic compounds. 1-Deuterioalkyl halides can be prepared by the exchange reaction of *S*-alkylsulfonium salts with deuterium oxide. The labelled alkyl halides can be released from the salt by thermal decomposition. The sulfonium salts of 4-phenylthiocyclohexane (**3.41**) have been used for this purpose. Thus, exchange of the hydrogen atoms of the methyl group of the methylsulfonium bromide in deuterium oxide containing 0.04 M sodium deuteroxide took place with minimal exchange at the C-2 and C-6 methylenes. [$^2$H$_3$]Methyl bromide was then obtained by decomposition of the salt at 160–170 °C. Other sulfonium and sulfoxonium salts have also been used for exchange reactions. Thus, heating trimethylsulfoxonium iodide in deuterium oxide containing a small amount of potassium carbonate gave [$^2H_9$]trimethylsulfoxonium iodide. [$^2H_3$]Methyl iodide and [$^2H_6$]dimethyl sulfoxide were obtained from this deuteriated salt by fractional distillation. The alkyl halides may also be obtained from the corresponding labelled alcohols by reaction with, for example, hydrogen iodide. Thus, [$^3H$]methyl iodide has been obtained from [$^3H$]methanol. Because of problems arising from the volatility and instability of [$^3H$]methyl iodide, crystalline [$^3H$]methyl sulfonate esters such as [$^3H$]methyl toluene-*p*-sulfonate or [$^3H$]methyl 4-nitrophenylsulfonate have been used.

The deuterium or tritium labels are stable to the conditions that have been employed for the quaternization of amines. Thus, quaternization of *N*-dimethylaminoethanol with [$^3H$]methyl iodide gave [*N-methyl-$^3H$*]-choline iodide. The replacement of an *N*-methyl group by an *N*-trideuteriomethyl group has been used in a number of situations to provide internal standards for gas chromatographic–mass spectrometric analysis. Deuteriated samples of the methylated xanthines, caffeine, theophylline and theobromine have been prepared as internal standards for analytical purposes. Replacement of the *N*-methyl groups by *N*-[$^2H_3$]methyl groups was achieved by methylation of synthetic precursors using [$^2H_3$]methyl iodide or [$^2H_6$]dimethyl sulfate (Scheme 3.7). For example, methylation of 6-aminouracil (**3.42**) with alkaline [$^2H_6$]dimethyl sulfate gave [$^2H_6$]-1,3-dimethyl-6-aminouracil (**3.43**). The imidazole ring of the purine was added *via* the 5-nitroso derivative followed by reduction and formylation to give [$^2H_6$]theophylline (**3.44**).

The alkylation of enolate anions with [$^2H_3$]methyl iodide also proceeds without loss of label. The steroid [4-$^2H_3$]methyltestosterone was prepared by alkylation of testosterone (**3.15**) in the presence of potassium *tert*-butoxide. There is also no significant loss of label when a Grignard reagent is prepared from [$^2H_3$]methyl iodide. Examples of the use of this are found in the preparation of labelled samples of mevalonic acid (see Chapter 4).

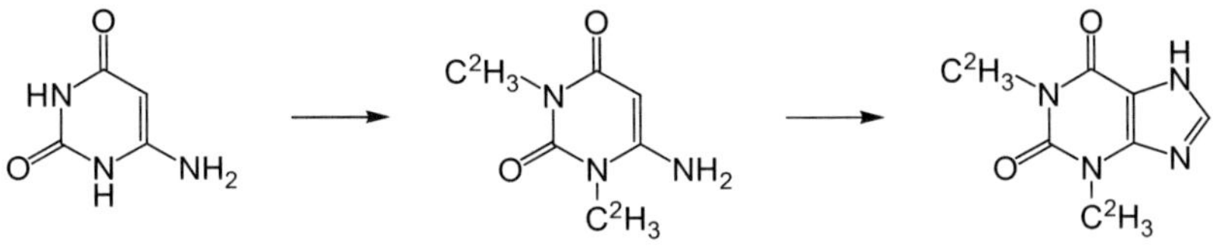

**3.42**          **3.43**          **3.44**

**Scheme 3.7**   The synthesis of [*methyl-$^2H_6$*]theophylline.

CHAPTER 4
# Stereochemical Aspects of Labelling with Hydrogen Isotopes

## 4.1 INTRODUCTION

Isotopic labelling using deuterium and tritium has played a major role in elucidating the stereochemistry of many chemical and biochemical reactions. The uses of labelled compounds, particularly in biochemical studies, have often depended not just on a knowledge of the relative stereochemistry of a label but also on a knowledge of the absolute stereochemistry of a chiral molecule. In this chapter we discuss examples of the stereospecific and enantiospecific labelling of methyl groups, methylenes, alkenes and alcohols with hydrogen isotopes. Although the examples are drawn mainly from the terpenoids and steroids and their biosynthetic precursors, where their preparation has been thoroughly explored, the methods have much wider applicability. Examples of the labelling of amino acids are discussed in Chapter 5.

In discussing the stereochemical outcome of reactions, a distinction can be made between those features such as steric hindrance which contribute to establishing the relative stereochemistry between centres and those features such as chiral auxiliaries which determine the absolute stereochemistry of a reaction at a centre. A labelled asymmetric centre may be created by asymmetric induction through the use of a chiral auxiliary, which may be removed from the structure once the chiral centre has been established. The labelled asymmetric centre may also arise because of the inherent asymmetry of the starting material.

The Organic Chemistry of Isotopic Labelling
By James R. Hanson
© James R. Hanson 2011
Published by the Royal Society of Chemistry, www.rsc.org

Since enzyme-catalysed reactions, by virtue of the structure of the enzyme, are almost always asymmetric, combinations of chemical and enzymatic steps are widely used in the preparation of stereospecifically labelled compounds for biochemical studies.

## 4.2 THE APPLICATION OF THE CAHN–INGOLD–PRELOG SEQUENCE RULES TO LABELLED COMPOUNDS

The *R/S* nomenclature based on the Cahn–Ingold–Prelog sequence rules provides a valuable way of describing the configuration of an asymmetric centre. The Cahn–Ingold–Prelog convention depends on ranking the atoms that are bonded to an asymmetric centre in terms of their atomic number. The higher the atomic number, the higher is the priority in the ranking. A tetrahedral asymmetric centre is then viewed with the atom of lowest rank at the back. If the sequence around the chiral centre which links the atom of highest priority with the next highest is clockwise, the centre has the *R* (*rectus* = right-handed) configuration. On the other hand, if the sequence is anti-clockwise, the centre has the *S* (*sinister* = left-handed) configuration. In situations where it is not possible to rank two substituents using the atoms directly bound to the asymmetric centre, the ranking is based on the next atoms moving outwards. If an atom is doubly bonded to a second atom, the priority system treats the first atom as though it were singly bonded to two of the second atom. When the asymmetry is created by isotopic substitution, the heavier isotope takes precedence over the lighter isotope.

The Cahn–Ingold–Prelog convention has been extended to the description of the geometry of alkenes. The substituents attached to the $sp^2$ carbon of the alkene are ranked. The *Z* (*zusammen* = together) isomer has the substituents of highest priority on the same side of the alkene whereas the *E* (*entgegen* = opposite) isomer has the groups on the opposite side. The face of an $sp^2$ centre such as the carbon of a carbonyl group or an alkene can also be described in terms of the sequence rules. The groups attached to the $sp^2$ centre are ranked and then the centre is viewed from the face which is to be described. If the priority sequence is clockwise, the face is known as the *re*-face, whereas if it is anti-clockwise, it is known as the *si*-face. These rules are particularly important in describing the interaction of labelled substrates with enzyme systems.

When a substrate is bound to an enzyme, the apparently symmetrical hydrogen atoms of a methylene that is attached to two dissimilar groups within the substrate become asymmetric by virtue of the asymmetry of the enzyme. If one of these hydrogens is stereospecifically replaced by deuterium or tritium, the unbound methylene then becomes asymmetric.

Using the Cahn–Ingold–Prelog sequence rule, these labelled centres become distinguishable as *R* or *S* centres. The hydrogen atoms whose labelling generates the chirality are known as the pro-*R* or pro-*S* atoms. This distinction does not just apply to hydrogen atoms. For example, the two carboxymethylene arms of citric acid when separately labelled become distinguishable. This distinction is particularly relevant when citric acid is enzyme bound in the citric acid cycle. This has already been refered to in Chapter 2. Although the *R/S* nomenclature is invaluable for describing the absolute stereochemistry of a centre, it is worth noting that the priority of the substituents that define the particular *R* or *S* designation of the centre may change as a consequence of reactions, *i.e.* an *R* centre may become an *S* centre not because of a reaction at that centre but because the composition of a substituent has changed.

## 4.3  THE STEREOCHEMISTRY OF REACTIONS USED IN LABELLING

The stereochemical consequences of some well-established reactions have made them particularly useful for labelling compounds with hydrogen isotopes. Before discussing some detailed examples, we will describe some underlying principles which contribute to this. First, there are those reactions in which the mechanism imposes clear stereochemical constraints. Second, there are reactions in which the stereochemical outcome is determined by the steric hindrance of a particular face of the molecule, which then directs attack of a reagent to the opposite face. Third, there are those reactions of a centre in which neighbouring group participation by an adjacent substituent facilitates a particular stereochemical outcome.

A bimolecular nucleophilic $S_N2$ substitution reaction proceeds with inversion of configuration. In the context of introducing a label, the nucleophile may be provided by a 'hydride' such as lithium aluminium deuteride or lithium borotritide and the leaving group may be the sulfonate of an alcohol. Thus the reduction of 5α-cholestane 3β-toluene-*p*-sulfonate (**4.1**) using lithium aluminium deuteride affords the 3α-deuterio steroid **4.2** (Scheme 4.1). When the corresponding 3-ketone was reduced first with lithium aluminium deuteride to form the 3α-deuterio-3β-alcohol and then the toluene-*p*-sulfonate of this was reduced with the deuteride, the product was [3-$^2H_2$]-5α-cholestane. It is also worth remembering that if a labelled alcohol with an undesired configuration has been generated by reduction of a ketone, the configuration may be inverted without loss of the label by the nucleophilic substitution of a sulfonate ester of the alcohol with an oxygen nucleophile such as an acetate or nitrite. The same result may be achieved by using the

**4.1**                                    **4.2**

**Scheme 4.1**   The reduction of 5α-cholestane 3β-toluene-*p*-sulfonate.

Mitsunobu reaction in which the alcohol is treated with triphenylpho-sphine, diethyl azodicarboxylate (DEAD) and a carboxylic acid such as benzoic acid.

An example of this labelling procedure is found in studies on fatty acid metabolism. The four epimeric C-9 and C-10 [$^3$H]stearic acids **4.3** were required in order to determine the stereochemistry of the enzymatic dehydrogenation of stearic acid to oleic acid (**4.4**). The epimers of the naturally occurring C-9 and C-10 alcohols were prepared by nucleo-philic substitution of the toluene-*p*-sulfonates. The labelled stearic acids were then prepared by tritiolysis of the toluene-*p*-sulfonates of the epi-meric pairs of C-9 and C-10 alcohols with lithium aluminium [$^3$H]hy-dride. The results of incubations with *Corynebacterium diphtheriae* were consistent with a stereospecific *cis* loss of hydrogen from C-9 and C-10 of stearic acid in the formation of oleic acid.

$$CH_3(CH_2)_7CH^2H.CH^2H.(CH_2)_7CO_2H$$
$$\phantom{CH_3(CH_2)_7}10\phantom{H.}9$$

**4.3**

cis

$$CH_3(CH_2)_7CH=CH(CH_2)_7CO_2H$$

**4.4**

Tritiated chiral alkanes were required as substrates for a methane monooxygenase from *Methylococcus capsulatus* (Scheme 4.2). [1-$^2$H]Butyraldehyde was prepared by oxidation of [1-$^2$H$_2$]butanol and then reduced with the stereospecific reagent *S*-Alpine-Borane$^®$ to give (*R*)-[1-$^2$H]butan-1-ol. This alcohol was then converted to its toluene-*p*-sulfonate (**4.5**) and treated with lithium triethyl[$^3$H]borane to give

**4.5**                                    **4.6**

**Scheme 4.2**   The synthesis of (*R*)-[1-$^2$H,$^3$H]butane.

$(R)$-[1-$^2$H,1-$^3$H]butane (**4.6**). Because of the way in which the sequence rule operates, although inversion occurs in the sequence, the centre is still described as an $R$ centre in the Cahn–Ingold–Prelog convention because the incoming tritium takes precedence over deuterium and hydrogen.

The well-established stereochemistry of addition reactions to alkynes and alkenes has led to a number of these being used in labelling sequences. The catalytic hydrogenation of alkynes and alkenes normally proceeds in a *syn* manner. However, catalytic deuteriation of an alkene over a palladium catalyst is not always a simple addition reaction. Allylic exchange reactions may also occur, giving polydeuteriated species. For example, catalytic deuteriation of cholest-2-ene has been reported to lead to the introduction of up to five deuterium atoms. Homogeneous catalysts may be more specific.

The stereochemistry of the enzymatic dehydrogenation of succinic acid to fumaric acid was established (Scheme 4.3) by using epimeric samples of [1,2-$^2$H$_2$]succinic acid which were prepared by the catalytic deuteriation of the corresponding unsaturated acids. Samples of fumaric (*trans*) acid (**4.7**) and maleic (*cis*) acid (**4.9**) were reduced with deuterium using a palladium on charcoal catalyst. The stereochemistry of the resultant samples of [1,2-$^2$H$_2$]succinic acid is shown in the structures **4.8** and **4.10**. Dideuteriated samples of fumaric acid were obtained only when the [1,2-$^2$H$_2$]succinic acid derived from maleic acid was subjected to enzymatic dehydrogenation. This corresponded to a *trans* elimination.

By using a partially poisoned Lindlar (Pd–CaCO$_3$–PbO) catalyst, the hydrogenation of an alkyne may be halted at the alkene stage affording a

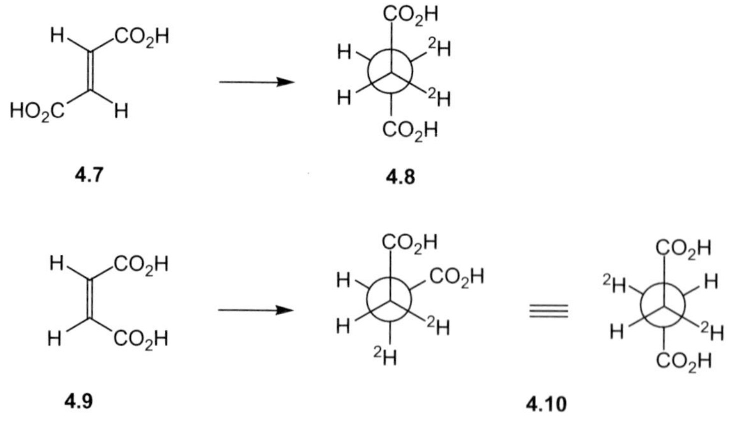

**Scheme 4.3**  The reduction of fumaric and maleic acids with deuterium.

*cis*-alkene. The reactivity of the alkene may be exploited in further labelling reactions. Diimide (diazene, $N_2H_2$) will reduce an alkyne to a *cis*-alkene. A useful reduction which proceeds in a *trans* manner uses triphenylphosphine and deuterium oxide or tritiated water. Although it is restricted to alkynes bearing substituents which activate the alkyne towards nucleophilic attack by the triphenylphosphine, it nevertheless has been particularly useful in the preparation of dimethyl [$^2H_2$]fumarate from dimethyl acetylenedicarboxylate. The labelled fumaric acid **4.7** has then been used as a substrate for the enzymatic preparation of (2*S*)-malic acid and L-aspartic acid (see Chapter 5).

Although they can be rather wasteful of label, the *cis* stereochemistry of hydroboration reactions has been used to introduce a label. The labelled borane is generated from deuteriated or tritiated sodium borohydride and boron trifluoride etherate. The hydroboronation of the steroidal 2-ene **4.11** followed by oxidation of the borane with alkaline hydrogen peroxide gives the hydration product in which there is a *cis* relationship between the 2α-deuterium or tritium and the 3α-hydroxyl group. The reaction is directed to the α-face of the alkene by the bulky β-oriented methyl group at C-10. Protonolysis (deuterolysis) of the borane with, for example, [*O*-$^2H$]propionic acid affords a reduction product. In some circumstances, this may be a more satisfactory method of stereospecifically introducing a label than catalytic reduction.

An important application of boranes in the preparation of labelled compounds is in the introduction of chirality through the reduction of carbonyl compounds by chiral boranes. These are discussed later in the context of the preparation of a chirally labelled primary alcohol (see Section 4.8).

The epoxidation of a double bond normally takes place from the less hindered face of an alkene without disturbing the relationship of the substituents on the double bond. An epoxide may also be generated stereospecifically by the internal *anti* displacement of a leaving group such as a toluene-*p*-sulfonate by an oxygen nucleophile derived from an adjacent alcohol. The reactive epoxide may then opened by the *anti* addition of a nucleophile such as a hydride, deuteride or tritide, creating a product with a defined stereochemical relationship between the label and a hydroxyl group on an adjacent carbon atom. Thus, epoxidation of the steroidal 2-ene **4.11** proceeds from the α-face to give the 2α,3α-epoxide. Reduction with lithium aluminium deuteride proceeded in a *trans* diaxial manner to give the 2β-deuterio-3α-alcohol. Note the difference in the stereochemistry of the label at C-2 compared with the hydroboronation of the 2-ene. Aziridines have also been used in similar sequences in the labelling of amino acids. There are a number of sequences where an epoxide has been used to protect an alkene. It may be removed with the regiospecific regeneration of the alkene *via* the action of zinc on the corresponding bromohydrin.

The stereochemical outcome of carbonyl addition reactions, and in particular the control exerted by the stereochemistry of neighbouring substituents, has played a widespread role in the use of these reactions in labelling sequences, particularly in chiral induction. When a carbonyl group is part of a rigid ring system, the trajectory of a nucleophile attacking the electron-deficient carbon will be governed by interactions with substituents such as methyl groups on the ring system. In the steroid series, the methyl groups at C-10 and C-13 (C-19 and C-18, respectively) lie on the β-face of the molecule and exert a significant stereochemical directing effect. Thus, reduction of steroidal 3-ketone **4.12** with sodium borodeuteride gave the equatorial 3α-deuterio-3β-alcohol, whereas reduction of an 11-ketone gave the axial 11α-deuterio-11β-alcohol. In both cases the reagent has approached the carbonyl group from the less hindered α-face of the molecule. On the other hand, reduction of the 11-ketone with lithium in [$^2$H$_3$]ammonia and [$O$-$^2$H]methanol gave the more stable equatorial 11β-deuterio-11α-alcohol. In this reduction, the outcome is determined by the relative stability of the alkoxide carbanions.

In a more flexible system in which the carbonyl group is free to rotate, more complicated conformational features need to be considered. The preferred conformation of the carbonyl group and the trajectory of the incoming reagents may be determined by the steric bulk of the adjacent substituents. In predicting the favoured conformation for a reaction, it is important to remember that many proceed by the initial co-ordinating of the reagent to the oxygen atom of the carbonyl group creating a bulky group out of the carbonyl oxygen. Where a cyclic six-membered transition state is involved in the reactions of a carbonyl group, the conformation of the substituents around the ring will determine the stereochemical outcome. However, to be of value for labelling purposes, the stereoselectivity of these reactions must be high.

The stereochemistry of the reactions of enolates is often determined by the stereochemistry of adjacent substituents. This can be particularly important in labelling processes involving enolate chemistry. The use of a number of chiral auxiliaries relies on these effects. Examples will be found both in this chapter and in the Chapter 5 on the synthesis of chiral amino acids.

The stereochemistry of electrocyclic reactions and sigmatropic rearrangements is governed by the rules of orbital symmetry. The Diels–Alder reaction and various sigmatropic rearrangements have been used not only to create centres of defined stereochemistry in labelled molecules but also to relay the stereochemistry at one centre to that of another. Examples can be found in the synthesis of the chiral methyl group (Section 4.6). In the sequel we will describe some topics that illustrate the ways in which these principles have been applied to labelling compounds.

## 4.4 LABELLED STEROIDS

The steroid hormones are major contributors to the biochemistry of the endocrine system and to the regulation of many natural processes. Isotopically labelled steroids have been used in developing methods for their identification and quantification and in studying their metabolism. Furthermore, the well-defined rigid framework of the steroids has provided the template for establishing the stereochemistry of many reactions, including a number of those involved in labelling compounds. The following sequences exemplify the preparation of labelled steroids for some of these studies. There are several general points which are illustrated by these examples. First, they reveal the influence of the structure of the steroid carbon skeleton in determining the stereochemistry of labelling reactions at particular centres. Second, they exemplify the strategy of temporarily introducing extra functional groups such as carbonyl groups and alkenes to facilitate labelling at particular sites, and third the use of 'back-exchange' to remove surplus labels.

Mass spectrometry, particularly when it is coupled with a powerful separation technique such as gas chromatography, is an important method for the identification of steroids and many other natural products. The mass spectrometric fragmentation pattern of saturated steroidal ketones is often very complex, involving hydrogen transfer from fairly distant parts of the molecule and not just the $\beta$-cleavage and $\gamma$-hydrogen transfer associated with simpler molecules. The elucidation of these fragmentation patterns has been explored with specifically deuteriated steroids. An example is provided by studies on variously deuteriated samples of 5$\alpha$-androstan-3-one (**4.12**). The general methods that were used involved exchange of the enolizable $\alpha$-positions to ketones, the $\gamma$-positions of $\alpha\beta$-unsaturated ketones, reduction (deuteriation) of unsaturated ketones and reduction of other carbonyl groups to methylenes. Thus, labelling at C-2 and C-4 was achieved through simple enolization and exchange. [1$\alpha$-$^2$H]-5$\alpha$-Androstan-3-one was prepared by catalytic deuteriation of 5$\alpha$-androst-1-en-3-one in which the addition was directed to the $\alpha$-face by the $\beta$-oriented C-10 methyl group. The deuterium at C-2 was removed by exchange. [6,6-$^2$H$_2$]-5$\alpha$-Androstan-3-one was prepared by base-catalysed enolization of androst-4-en-3-one to give [2,2,4,6, 6-$^2$H$_5$]androst-4-en-3-one. The double bond was reduced with lithium and ammonia and the desired 6,6-$^2$H$_2$ derivative was then obtained by back-exchanging the deuterium atoms to remove those attached to C-2 and C-4. The 5α-$^2$H derivative was obtained from the readily available 3β-acet-oxyandrost-5-en-7-one. Catalytic deuteriation from the α-face and base-catalysed back-exchange at C-6 followed by hydrolysis of the 3β-acetate gave 3β-hydroxy[5α-$^2$H]androstan-7-one (**4.13**). The carbonyl group which had facilitated the introduction of the deuterium label at the C-5α-position was then removed by a Wolff–Kishner reduction and the 3-ketone was generated by oxidation. Deuterium was regioselectively introduced on ring C in a similar manner by making use of C-11 and C-12 ketones as activating and directing groups. Mass spectrometric studies revealed the role of 'itinerant' hydrogen atoms in creating particular fragment ions.

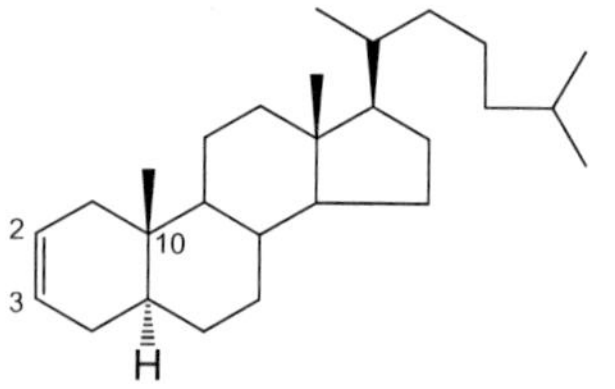

**4.11** 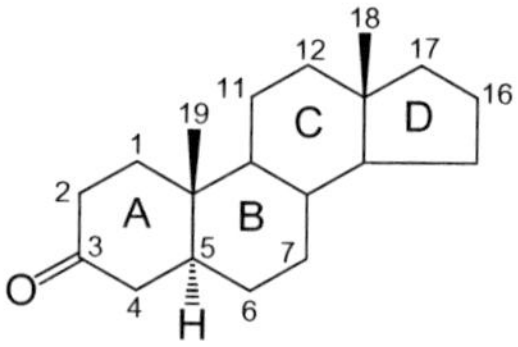　　　　　　　　**4.12** 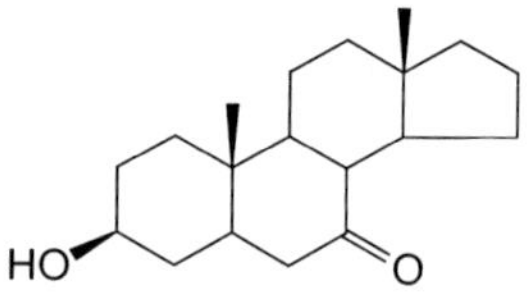　　　　　　　　**4.13**

Studies of the stereochemistry in the mechanism of estrogen bio-synthesis from androst-4-ene-3,17-dione (**4.14** → **4.15**) (Scheme 4.4) have required stereospecifically labelled steroids. The reaction, which is catalysed by the enzyme system aromatase, involves the loss of the C-10 methyl group (C-19) as formic acid and a hydrogen atom from both C-1 and C-2. It is a three-step oxidative sequence involving hydroxylation at C-19, oxidation to a C-19 aldehyde and aromatization with the loss of C-19 and the hydrogen atoms. Inhibition of this sequence is important

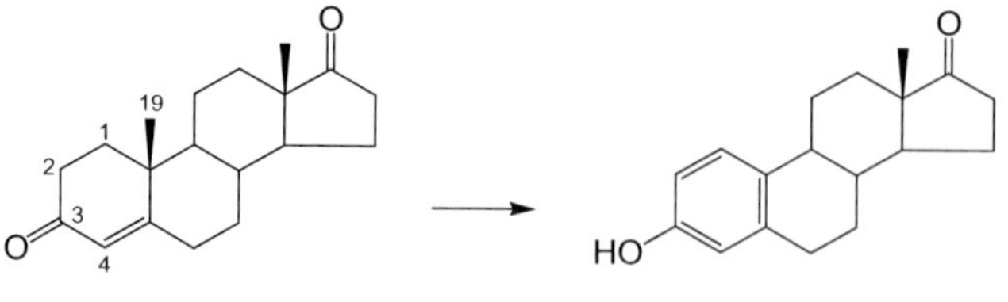

**4.14**　　　　　　　　　**4.15**

**Scheme 4.4**　Estrogen biosynthesis from androst-4-ene-3,17-dione.

in the treatment of breast cancer. Labelled steroids for these studies were prepared by the following sequences.

The homogeneous deuteriation of androsta-1,4-diene-3,17-dione (**4.16**) with the catalyst tris(triphenylphosphine)chlororhodium proceeds from the α-face to gives a good yield of [1α,2α-$^2$H$_2$]androst-4-ene-3,17-dione. The selective palladium-catalysed heterogeneous deuteriation of the dienone proceeded in poor yield and some addition took place from the β-face. Back-exchange removed the C-2 labels to give the C-1-labelled substrate. The stereochemistry of labelling at C-1 was established by bromination at C-2 to give a 2α-bromo compound and then measuring the remaining proton–proton coupling constants in the NMR spectrum.

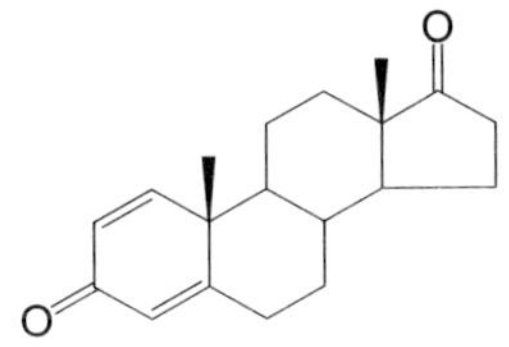

**4.16**

The stereochemistry of hydrogen elimination from C-2 was established by stereospecifically labelling the C-2α and 2β-hydrogen atoms. The labelling procedures (Scheme 4.5) were based on the reactions of the 2-ene **4.17**. Epoxidation formed the 2α,3α-epoxide which underwent a stereospecific reduction with lithium aluminium tritide to give the [2β-$^3$H]androstan-3α,5α,17β-triol (**4.18**). Acetylation, dehydration, mild hydrolysis and oxidation led to [2β-$^3$H]androst-4-ene-3,17-dione (**4.14**). The 2α-$^3$H epimer was obtained by a tritiohydroboronation of the 2-ene, which was then converted to [2α-$^3$H]androst-4-ene-3,17-dione (**4.14**). NMR studies showed that the reduction of a 19-aldehyde with sodium [$^2$H]borohydride gave predominantly the 19($S$)-$^2$H-labelled alcohol in a reduction which was governed by the steric approach of the reagent to C-19. The outcome of these labelling studies was that in placental estrogen biosynthesis, the C-1β and C-2β hydrogen atoms were removed

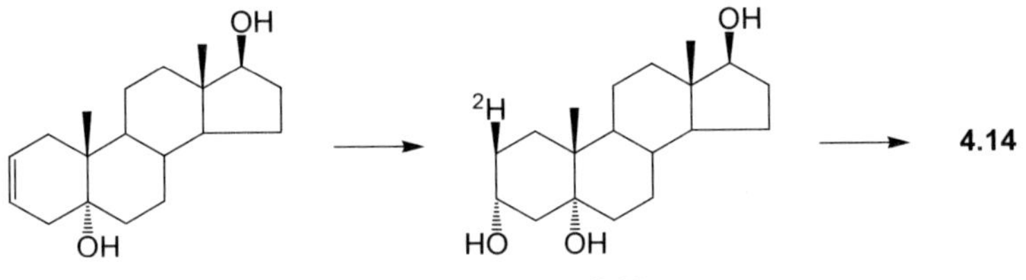

**4.17**          **4.18**          **4.14**

**Scheme 4.5**   The synthesis of [2β-$^3$H]androst-4-ene-3,17-dione.

and that in the conversion of the 19-alcohol to the aldehyde, there is the loss of the pro-*R* 19-hydrogen atom.

Whereas the previous examples were concentrated on introducing one deuterium label into the substrate, multiply labelled derivatives are often used in quantitative analysis in order to avoid complications arising from the presence of naturally occurring isotopes such as carbon-13. Multiply labelled steroids are used in order to bring the molecular ion of the steroidal standard to a different value from that of the cluster associated with the natural steroid. The formation of $[M + H]^+$ ions and the presence of carbon-13 can complicate the quantitative analysis of the molecular ion. The introduction of three deuterium labels into the standard is considered to be an optimum number in this context.

The detection of abnormal testosterone:17-epitestosterone ratios (17β-OH:17α-OH) in urine samples has been used as an indication of the abuse of the anabolic steroid testosterone (**4.21**) by athletes. A simple synthesis (Scheme 4.6) of triply labelled testosterone and 17-epitestosterone for analytical standards was based on the labelling of C-16 and C-17 by making use of the reactions of a C-17 ketone. This partial synthesis indicates the need for the careful selection of protecting groups in the specific labelling of a multifunctional molecule. In order to achieve the labelling of ring D, ring A of the steroid was protected by conversion of dehydroisoandrosterone to the 6β-methoxy-3,5-cyclo-steroid **4.19**. A one-pot exchange and reduction of the 17-ketone with sodium methoxide in [O-$^2$H]methanol followed by the addition of excess sodium gave the [16,16,17-$^2$H$_3$]-17β-alcohol. The preparation of the

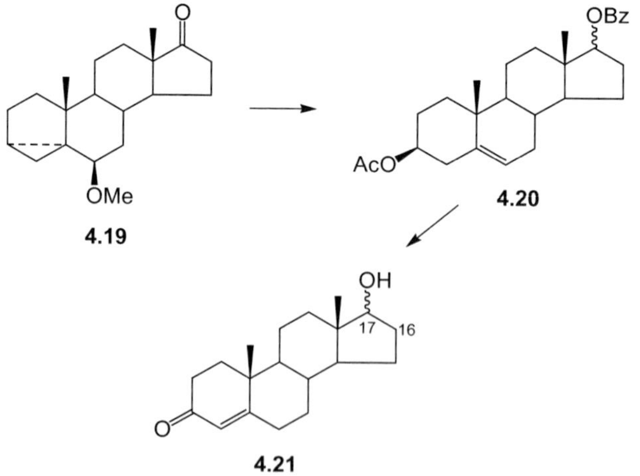

**Scheme 4.6**   The synthesis of [16,16,17-$^2$H$_3$]testosterone.

labelled 17-epitestosterone required inversion of the 17β-alcohol. Hence this alcohol was converted to its toluene-*p*-sulfonate, which was then displaced with potassium nitrite in dimethylformamide to give the 17α-alcohol without the loss of label at either C-16 or C-17. The generation of the unsaturated ketone on ring A of testosterone and 17-epitestosterone involved the selective oxidation of C-3 while the C-17 alcohols were left untouched. Consequently, the 17-alcohols were protected as their bulky benzoate esters. The 6β-methoxy-3,5-cyclopropane which had protected ring A was then cleaved and ring A was converted to the 3β-acetoxy-5-ene **4.20**. The more reactive 3β-acetate ester was then hydrolysed. Oppenauer oxidation with aluminium isopropoxide and 1-methyl-4-piperidone as the hydrogen acceptor gave the unsaturated 4-en-3-one typical of the hormonal steroid. Finally, hydrolysis of the 17-benzoate afforded [16,16,17-$^2$H$_3$]testosterone and epitestosterone (**4.21**). The labelled testosterone was produced without the inversion steps.

## 4.5   LABELLED GIBBERELLIN PLANT HORMONES

Many stages of plant growth and development are regulated by the gibberellin plant growth hormones. These substances occur in plant tissues at very low concentrations. The deuteriation of the gibberellins was undertaken in order to investigate their metabolic relationships and to facilitate the mass spectrometric identification and quantification of the small amounts of these hormones that are found in Nature. Many of the methods that have been used to label the gibberellins parallel those found with the steroids, but others, some of which will be described here, reveal different stereochemical features in labelling multifunctional molecules. The chemistry of gibberellic acid (**4.22**) is complicated by the facile acid-catalysed aromatization of ring A and, under more vigorous conditions, a Wagner–Meerwein rearrangement of rings C and D. The lactone and double bond of ring A also undergo rearrangement in the presence of base.

**4.22**          **4.23**

The functionalization of ring A of these hormones is intimately connected with their metabolism and biological activity. Gibberellic acid and the 13-desoxygibberellin mixture of gibberellins $A_4$ and $A_7$ which are produced by fermentation provided the most readily available starting materials. The 16,17-double bond can be protected as its epoxide. Catalytic deuteriation and tritiation of the 1,2-double bond proceeded from the less hindered β-face to give the 1β,2β-dideuterio product. This was then followed by a deprotection sequence in which the epoxide was converted to a bromohydrin. Reductive elimination with zinc dust regenerated the alkene of gibberellin $A_1$ (1,2-dihydrogibberellic acid) (**4.23**). This route has also been used to prepare tritiated material and to prepare the 13-desoxy analogue, [$^3$H]gibberellin $A_4$. Conjugate reduction of the 1-en-3-one **4.24** with sodium borodeuteride in methanol in the presence of CuCl gave mainly [1β,3β-$^2$H$_2$]-3-epigibberellin $A_4$ (**4.25**). Conjugate reduction in the presence of the deuteriated solvent MeO$^2$H afforded the C-2-labelled gibberellin, but this reduction, which proceeded through the enol, was less stereospecific. Hydrogenolysis of a 2β-iodo-3-ketone with a tributylstannane deuteride gave [2α-$^2$H]gibberellin $A_4$ 3-ketone. This group of reactions show how an unsaturated ketone can be used in a labelling sequence by making use of conjugate addition reactions.

**4.24**                                        **4.25**

3β-Hydroxygibberellins with a saturated ring A, undergo an isomerization to an epimer *via* a retro-aldol–aldol rearrangement reaction initiated by the lactone carbonyl group. This affords an aldehyde **4.26** as an intermediate which re-cyclizes to give the more stable 3α-alcohol with an equatorial hydroxyl group. If this epimerization is carried out in a deuteriated medium, exchange takes place at C-2 *via* the enolate of the transient aldehyde **4.26**. Exchange also occurs at C-6, which is the other enolizable position. Thus, the reaction of gibberellin $A_4$ methyl ester with sodium methoxide in [$O$-$^2$H]methanol gave mainly (2,2,6-$^2$H$_3$)-3-epigibberellin $A_4$ methyl ester. These labelled gibberellins were then converted to other gibberellins such as gibberellin $A_9$. This illustrates the point that any reaction which involves an enolization or a retro-aldol–aldol sequence has the potential for temporarily exposing parts of a molecule for use in labelling.

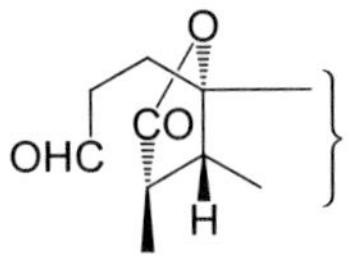

**4.26**

## 4.6  THE CHIRAL METHYL GROUP

If a carbon atom is covalently bound to the three isotopes of hydrogen and a fourth group X such as a carboxyl group, the methyl group is chiral. There are a number of enzyme-catalysed reactions in which methyl groups are generated from, or converted into, methylene groups. One example (Scheme 4.7) is the isomerization of isopentenyl diphosphate (**4.27**) (PP = diphosphate) to dimethylallyl diphosphate (**4.28**) in isoprenoid biosynthesis, and another is the condensation of acetyl coenzyme A with oxaloacetic acid (**4.29**) to form citric acid (**4.30**) in the tricarboxylic acid cycle. In the case of the former, when the hydrogens on the methylene of the alkene are individually labelled with deuterium and tritium, stereospecific protonation by an enzyme will lead to a chiral methyl group. Its chirality will reflect the face of the alkene that has been protonated. In the case of the latter, when a hydrogen atom is abstracted from the acetyl coenzyme A by the enzyme to generate a carbanion, the new carbon–carbon bond of the citric acid may be formed on the same or the opposite face, *i.e.* the enzyme-catalysed reaction may take place with retention or inversion of configuration of the methyl group. The solution to these and related problems of enzyme stereochemistry

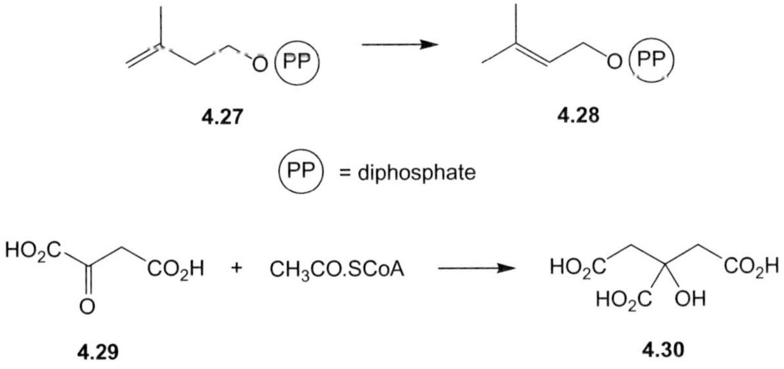

**4.27**                                **4.28**

$(PP)$ = diphosphate

**4.29**                                               **4.30**

**Scheme 4.7**   Some enzymatic processes involving a chiral methyl group.

required the synthesis of acetic acid possessing a chiral methyl group of known absolute configuration.

The methods that have been used to create the chiral methyl group of acetic acid have involved the sequential stereospecific addition of the isotopes of hydrogen to an alkyne to convert the sp-carbon to an sp$^3$-carbon. The principle underlying the Cornforth synthesis (Scheme 4.8) of chiral acetic acid was to prepare stereospecifically labelled 1-phenyl-ethanol [PhCH(OH)Me] and to resolve it into the 1$R$ and 1$S$ enantiomers (**4.34** and **4.35**), whose configuration was known from other work. The method of synthesis of the 1-phenylethanol meant that the stereochemistry of the secondary alcohol reflected the stereochemistry of the hydrogen isotopes on the methyl group (C-2). Deuteriophenylacetylene was prepared by decomposing the corresponding alkynyl Grignard reagent with deuterium oxide. The *cis* addition of hydrogen to the alkyne with diimide (diazene, N$_2$H$_2$) gave *cis*-1-phenyl [2-$^2$H]ethene (**4.31**). This was epoxidized with perbenzoic acid and the racemic mixture of epoxides, **4.32** and **4.33**, was reduced with lithium [$^3$H]borohydride. In this

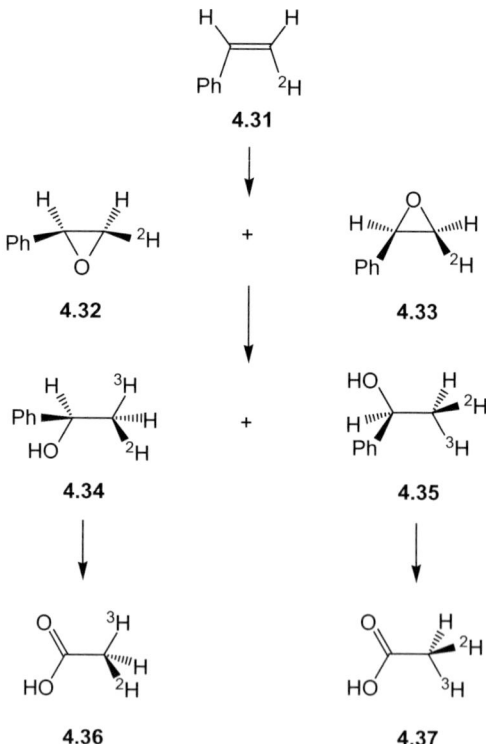

**Scheme 4.8**   The Cornforth synthesis of chiral acetic acid.

reduction, the incoming tritium stereospecifically displaced the epoxide oxygen at C-2 with inversion of configuration. The racemic mixture of C-1 alcohols, **4.34** and **4.35**, was then resolved through the brucine phthalates to give the individual (1*R*,2*R*)- and (1*S*, 2*S*)-[2-$^2$H$_1$, $^3$H$_1$]phenylethanols. The configuration of these were known because of prior correlations between optically active samples of 1-phenylethanol and mandelic and lactic acids. The separate 1-phenylethanols were then oxidized to the (*R*)- and (*S*)-acetophenones and subjected to a Baeyer–Villiger oxidation to give the phenylacetates. Hydrolysis of the acetates afforded the chiral acetic acids **4.36** and **4.37**.

One of the driving forces behind this synthesis of a chiral methyl group of known absolute stereochemistry was to develop a method for determining the configuration of a methyl group formed by an sp$^2$ → sp$^3$ conversion in the biosynthetic isomerization of isopentenyl diphosphate (**4.27**) to dimethylallyl diphosphate (**4.28**). Subsequently, the stereochemistry of a number of important enzymatic processes have been established using these labelled samples.

The principles which underlie the method (Scheme 4.9) for determining the stereochemistry of the chiral methyl group involve the isotope effects in the conversion of acetic acid to 2(*S*)-malic acid (**4.38**) and its subsequent dehydration to fumaric acid (**4.39**). Isotope effects would favour the retention of tritium and to a lesser extent deuterium over hydrogen. The dehydration of the 2(*S*)-malic acid takes place with *trans* stereospecificity. The results with the acetate showed that the enzymatic formation of 2(*S*)-malic acid from acetic acid (**4.36**) and glyoxalate by malate synthase proceeded with inversion of configuration of the acetic acid. In the dehydration of the 2(*S*)-malic acid to fumaric acid (**4.39**), there was a smaller loss of tritium from the (*R*)-[$^3$H$_1$,$^2$H$_1$]-acetic acid (**4.36**) compared with the (*S*)-[$^3$H$_1$,$^2$H$_1$]acetic acid (**4.37**). In understanding how this works, it is helpful to make a model and to remember that the carboxyl groups and the hydroxyl group of the 2(*S*)-malic acid lock it in a particular conformation within the enzyme.

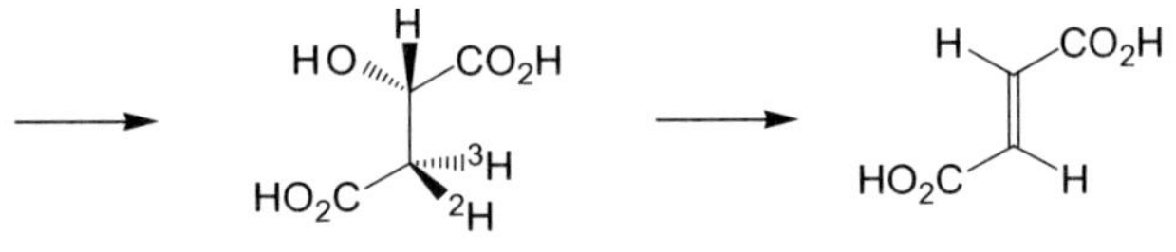

**Scheme 4.9**  The determination of the stereochemistry of the chiral methyl group.

   The well-defined stereospecificity of sigmatropic rearrangements have played an important role in two other syntheses of chiral acetic acid (Scheme 4.10). A synthesis of chiral acetic acid which was devised by Arigoni used an intramolecular chirality transfer reaction employing the cyclization of a doubly labelled chiral ether (**4.41**). The reaction depends on the transfer of two hydrogen atoms to an alkyne, the first by an ene reaction and the second by the elimination of methyl formate from a chiral methoxymethyl ether which was accompanied by the internal hydrogen rearrangement. A propargylic alcohol (**4.40**) which formed the starting material was resolved and converted to the doubly labelled species **4.41** by reaction with deuteriated chloromethyl methyl ether and exchange of the acetylenic hydrogen with tritiated water. The ene reaction generated the stereospecifically labelled methylenecyclopentene **4.42**. The 1,5 deuterium shift which accompanied the elimination of methyl formate from the chiral methoxymethyl ether then created the chiral methyl group in **4.43**. Vigorous chromic acid oxidation generated the chiral acetic acid **4.36**.

   A thermal sigmatropic [1,5]-hydrogen migration has been used in another chirality transfer sequence to create a chiral methyl group. The regiospecifically labelled diene **4.44** was prepared by an organometallic addition to an alkyne derived from phenylacetylene and a coupling reaction with a chiral iodoalkene to introduce the chiral side-chain. The thermal [1,5]-rearrangement of **4.44** then generated the chiral methyl group in **4.45**. The oxidative degradation of the diene by ozonolysis and a Baeyer–Villiger oxidation produced the chiral acetic acid **4.37** without racemization.

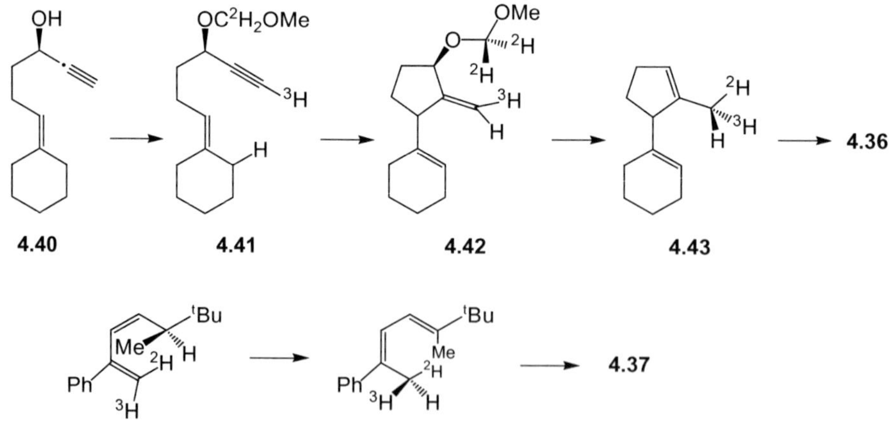

**4.40**           **4.41**                **4.42**                **4.43**                        4.36

**4.44**                 **4.45**           4.37

**Scheme 4.10**   Sigmatropic rearrangements in the synthesis of chiral acetic acid.

## 4.7  STEREOSPECIFICALLY LABELLED MEVALONIC ACID

Mevalonic acid is a key intermediate in the biosynthesis of terpenoids and steroids, particularly in mammals and fungi. Decarboxylation of the 3*R* enantiomer of the 5-diphosphate **4.46** affords isopentenyl diphosphate (**4.27**), which provides the characteristic isoprene building block. Another pathway based on 1-deoxy-D-xylulose (see Section 4.9) has been found in bacteria and plants. Variously labelled forms of mevalonic acid have been used to answer questions of isoprenoid biosynthesis. In Chapter 2, we considered the synthesis of carbon-labelled mevalonates which have been used in the identification of the constituent isoprene units of terpenoids. Some of the biosynthetic reactions such as cyclizations are accompanied by skeletal and hydrogen atom rearrangements. Alkenes are formed at particular stages in these biosyntheses with the stereospecific loss of a hydrogen atom. The later stages of many terpenoid and steroid biosyntheses involve hydroxylations and related oxidative processes. These biosynthetic reactions may proceed with retention or inversion of configuration. Both carbon-labelled and stereospecifically deuteriated and tritiated mevalonates have played a role in answering these questions.

**4.46**

The hydrogen atoms from C-2, C-4 and C-5 of mevalonic acid participate in many biosynthetic reactions. They have been stereospecifically labelled in several ways. Stereospecific labelling with deuterium and tritium at C-4 was based on the *trans* stereochemistry of the reduction of epoxides. The stereochemistry of labelling at C-4 was then transferred to C-2 by interchanging the arm containing the C-1 carboxyl with the C-5 primary alcohol; in one synthesis (Scheme 4.11), the unsaturated acid **4.47** was separated into its *cis* and *trans* isomers. The alkenes were separately converted to their epoxides. Thus the racemic mixture of the epoxides **4.48** and **4.49** was obtained from the *trans* isomer of **4.47**. Reduction of the epoxides with lithium borodeuteride or borotritide proceeded in a *trans* manner to give a mixture of **4.50** and its enantiomer **4.51**, which could be converted to mevalonic acid by removal of the amide protecting group. The label which was introduced at C-4 by this reduction has a specific *anti* geometric relationship to the new tertiary alcohol at C-3. Since only the 3*R* enantiomer **4.50** of the C-3 tertiary alcohol is used in the biosynthesis of isopentenyl diphosphate from

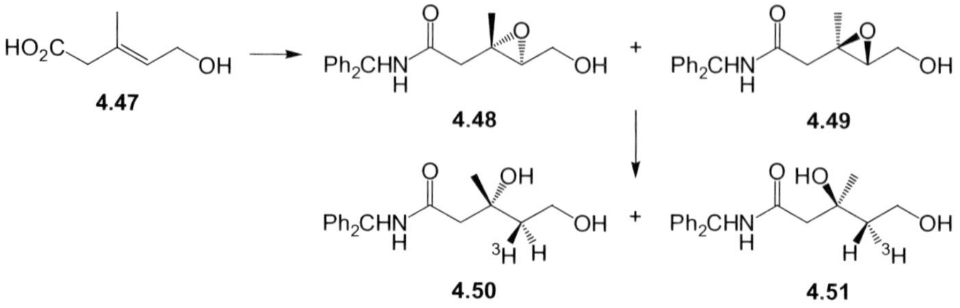

**Scheme 4.11**   The synthesis of mevalonic acid bearing a chiral label at C-4.

mevalonic acid, this is effectively a synthesis of the 'natural' mevalonic acid which is chirally labelled at C-4. The C-4 and C-2 positions in mevalonic acid were interchanged by a sequence involving the oxidation of the C-5 primary alcohol and a reduction of the C-1 carboxylic acid. Hence this provided a method for stereospecifically labelling C-2. Although there are potential hazards in making the assumption concerning the role of the chirality at C-3, it is worth remembering that in general an enzyme system will discriminate between two enantiomers. On occasions, a radiochemical synthesis that leads to a racemate can be used with care to explore a chiral process.

Another synthesis (Scheme 4.12) of [2- and 4-$^2$H]mevalonates was also based on the stereospecific reduction of chiral epoxides. The labelled *Z*-unsaturated ester **4.53** was prepared by the conjugate addition of dimethylcopperlithium to the acetylenic ester **4.52** followed by quenching with deuterium oxide. Reduction of the ester with lithium aluminium hydride gave an allylic alcohol. Asymmetric epoxidation of the allylic alcohol with *tert*-butyl hydroperoxide, titanium isopropoxide and (+)-diethyl tartrate (Sharpless conditions) gave the (2*S*,3*R*)-epoxy alcohol **4.54**. This was reduced with lithium aluminium hydride to give **4.55**. The unprotected primary alcohol was then oxidized with alkaline potassium permanganate before acid hydrolysis exposed the other hydroxyl group, leading to the formation of the (2*R*,3*R*)-mevalonolactone **4.56**. The diastereomeric 2*S*-labelled mevalonate was obtained by quenching the methylcuprate adduct with water and reducing the epoxide with lithium aluminium deuteride. The procedure for introducing the label at C-4 utilized the differing reactivity of ethoxyethyl and acetoxyl protecting groups for the primary alcohol in order to interchange C-1 and C-5.

The hydrogen atoms at C-5 of mevalonic acid are also involved in various enzymatic steps. Enzymatic reduction of the C-5 aldehyde, mevaldic acid with a pig or rat liver mevaldate reductase and

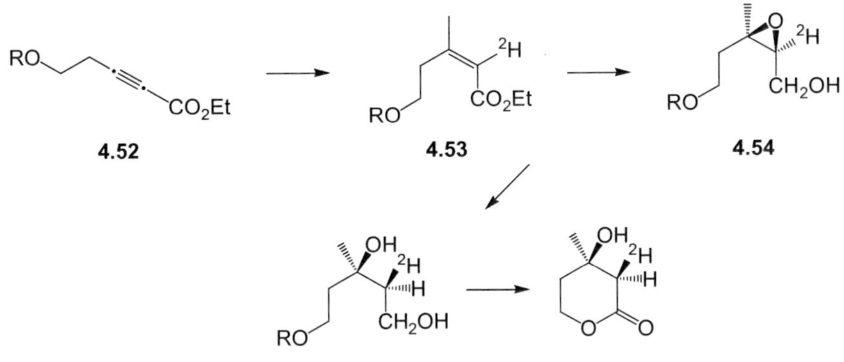

**Scheme 4.12** The synthesis of mevalonic acid bearing a chiral label at C-2.

$[4(R)-4-^3H]$NADPH as the source of the label afforded the chiral $[5(R)-5-^3H]$mevalonic acid **4.57**. The $5(S)-5-^3H$ isomer has been obtained by reducing $[5-^3H]$ mevaldic acid with the same enzyme and NADPH. The tritiated aldehyde of mevaldic acid was prepared by condensing methyl $[^3H]$formate and acetone followed by treatment with methanolic hydrogen chloride. The resultant $[4-^3H]-4,4$-dimethoxybutan-2-one was converted to the labelled mevaldic acid.

Another chemo-enzymatic approach to $[3(RS),5(S)-5-^3H]$ mevalonolactone employed the following strategy. In primary alcohols generated from aldehydes by reduction with liver alcohol dehydrogenase, the 1-pro-*R* hydrogen atom comes from the coenzyme and the 1-pro-*S* hydrogen atom is derived from the substrate. 3-Methyl[1-$^3H$]but-3-en-1-al formed the starting material. Since this aldehyde readily isomerizes to the conjugated $\alpha,\beta$-unsaturated aldehyde, its preparation in a labelled form had to be conducted under carefully controlled conditions. It was reduced enzymatically with the liver alcohol dehydrogenase and reduced nicotinamide adenine dinucleotide. The 1(S)-[1-$^3H$]-3-methylbut-3-en-1-ol was purified and converted to the (3RS)-mevalonic acid *via* the bromohydrin **4.58**, The bromine was displaced with potassium cyanide and the cyanide hydrolysed with sodium hydroxide to form the carboxylic acid of mevalonic acid. This procedure has also been used to make deuteriated material.

## 4.8 STEREOSPECIFICALLY LABELLED ISOPENTENYL DIPHOSPHATE

Two $C_5$ building blocks, isopentenyl diphosphate (**4.27**) and dimethylallyl diphosphate (**4.28**), that are derived from mevalonic acid, combine to form the polyisoprenoid chains that are the precursors of the cyclic terpenoids and steroids. Dimethylallyl diphosphate initiates the head-to-tail chain elongation by an alkylation of isopentenyl diphosphate generating an allylic diphosphate chain. Consequently, labelled forms of these $C_5$ building blocks have been prepared in order to study various aspects of the action of the prenyl synthases. The stereospecific labelling of isopentenyl diphosphate illustrates a different range of methodologies when compared with those that have been used to label mevalonic acid. Some of these will be described in this section.

The 1(*R*)- and 1(*S*)-[1-$^2$H$_1$]isopentenyl diphosphates were prepared from 3-methylbut-3-enoic acid by reduction of the carboxylic acid with lithium aluminium deuteride. Careful oxidation of the deuteriated primary alcohol to avoid double bond isomerization gave the labelled 3-methylbut-3-enal, which was reduced with a chiral borane. Hydroboration of a chiral terpene, α-pinene, which is available in both enantiomeric forms, with a bulky borane, 9-borabicyclononane (9BBN), gives the chiral *R*- or *S*-Alpine-Borane®, *e.g.* **4.59**. When these are used to reduce a carbonyl compound, the hydrogen that is transferred is derived from the original borane *via* the chiral α-pinene. It is therefore delivered in a chiral manner. Labelled Alpine-Boranes® can be prepared from the corresponding [$^2$H]- or [$^3$H]9BBN. These were used to give the stereospecifically labelled isopent-3-en-1-ols. The diphosphate was introduced by displacement of the toluene-*p*-sulfonate with tris(tetra-*n*-butylammonium) hydrogendiphosphate in acetonitrile in a reaction which proceeded with inversion of configuration at C-1. A similar procedure has been used to prepare (*S*)-[1-$^2$H]farnesol for a study of the farnesyltransferase reaction in yeast. Methyl farnesoate was reduced with lithium aluminium deuteride to [1-$^2$H$_2$]farnesol. This was then oxidized with manganese dioxide to [1-$^2$H]farnesal. The (*S*)-[1$^2$H]farnesol was obtained from this aldehyde by reduction with *R*-Alpine-Borane®.

The C-2 stereospecifically labelled isopentenyl diphosphates were prepared from the L- or D-glyceraldehydes (Scheme 4.13). The strategy depended on the reduction of a chiral epoxide in which the chirality was created using the chiral secondary alcohol of the glyceraldehyde. The acetal **4.60** derived from L-glyceraldehyde was converted *via* a methyl ketone and a Wittig reaction to the alkene **4.61**. The primary alcohol in **4.61** was then converted to a toluene-*p*-sulfonate and, under basic conditions, ring closure by the secondary alcohol occurred to generate an epoxide. Reduction of the epoxide with sodium cyanoborodeuteride in the presence of boron trifluoride etherate produced the enantiospecifically deuteriated alcohol **4.62**, which was then converted to the diphosphate. The enantiomer was prepared from D-glyceraldehyde.

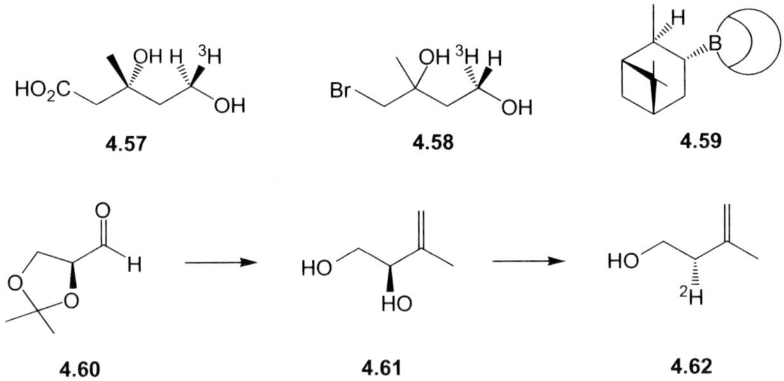

**Scheme 4.13**   The stereospecific labelling of isopentenol at C-2.

The alkene hydrogens were stereospecifically labelled by deuterolysis of bromoalkenes. Bromine was added to isopent-3-en-1-ol and elimination of HBr with methanolic potassium hydroxide then gave a separable mixture (4:1) of (*E*)- and (*Z*)-4-bromo-3-methylbut-3-en-1-ols. The hydroxyl groups of the isomers were protected as their *tert*-butyldimethylsilyl derivatives and the bromine atoms were then replaced stereospecifically with deuterium by treatment with *tert*-butyllithium and deuteriated trifluoroacetic acid to give the deuteriated isopentenols from which the diphosphate was prepared.

## 4.9   LABELLING OF 1-DEOXY-D-XYLULOSE

1-Deoxy-D-xylulose-5-phosphate (**4.66**, R = PO$_3$H$_2$), is an intermediate in the biosynthesis of pyridoxal, thiamin and isopentenyl diphosphate in the non-mevalonoid pathway of isoprenoid biosynthesis. This pathway has attracted interest in recent years because, in contrast to the mevalonate pathway, it is a major pathway for the formation of isoprenoids in bacteria and in the chloroplasts of higher plants. 1-Deoxy-D-xylulose is biosynthesized from pyruvic acid and glyceraldehyde monophosphate. The synthesis of labelled forms of 1-deoxy-D-xylulose reveals an interplay between chemical and enzymatic methods that characterize many modern syntheses of labelled compounds.

[5-$^3$H$_2$]-1-Deoxy-D-xylulose (**4.66**, R = H) has been prepared (Scheme 4.14) from dimethyl 2,3-isopropylidene-D-tartrate (**4.63**). Regiospecific hydrolysis using a pig liver esterase enabled a distinction to be made between the two ester groups and afforded the chiral monoester **4.64**. The ester was reduced with lithium triethylborodeuteride to give the deuteriated primary alcohol and the carboxyl group was then converted

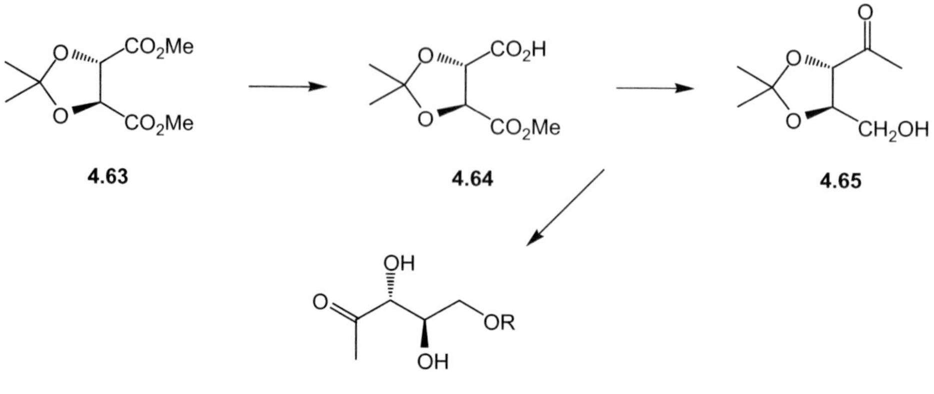

**Scheme 4.14**  The labelling of 1-deoxy-D-xylulose.

to the methyl ketone **4.65** by reaction with methyllithium. Removal of the isopropylidene protecting group gave [5-$^2$H$_2$]-1-deoxy-D-xylulose (**4.66**, R = H).

Another approach was based on the biosynthesis of 1-deoxy-D-xylulose. A recombinant 1-deoxy-D-xylulose synthase from *Bacillus subtilis* has been isolated. This enzyme system combines pyruvic acid and glyceraldehyde 3-phosphate to give the pentose phosphate **4.66** (R = PO$_3$H$_2$), with the loss of carbon dioxide from the pyruvic acid. Given the ease of making variously labelled forms of these substrates, this enzyme system made the synthesis of different isotopomers of 1-deoxy-D-xylulose 5-phosphate possible. Although it was exemplified by the preparation of samples labelled with carbon-13 on each of the five carbons, the availability of substrates bearing a hydrogen isotope means that this biosynthetic method could be used for the preparation of stereospecifically deuteriated and tritiated material. Bearing in mind the substrate flexibility of various aldolases, this approach is a potentially useful, more general strategy for the preparation of labelled carbohydrates.

## 4.10   THE STEREOCHEMISTRY OF HYDROGEN TRANSFER FROM THE NICOTINAMIDE COENZYMES

The determination of the stereochemistry of the enzymatic hydrogen transfer to the pyridine nucleotides (NAD$^+$ and NADP$^+$) at C-4 made use of the optical activity arising from the chiral centre which can be created when a hydrogen of a methylene is stereospecifically replaced by deuterium. The two hydrogen atoms, H$_A$ and H$_B$, attached to C-4 of the dihydronicotinamide are distinct (Scheme 4.15). The pyridine

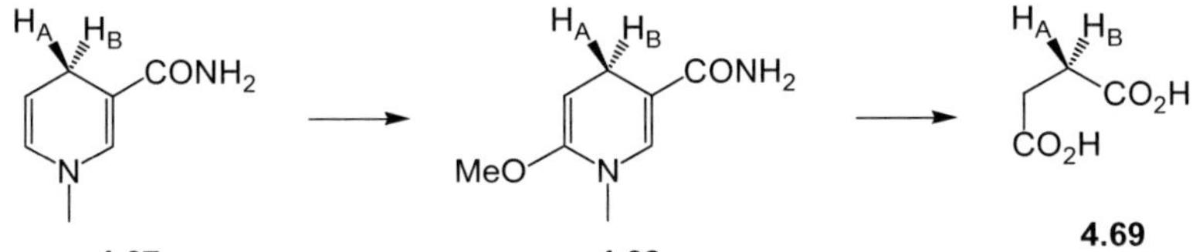

**Scheme 4.15** The degradation of NADH to succinic acid.

nucleotide-dependent enzymes can be divided into two classes according to their stereochemical specificity in utilizing either $H_A$ or $H_B$ in redox reactions. The requirement was to establish the stereochemistry of $H_A$ and $H_B$. Two specimens of [4-$^2$H]NADH (**4.67**) were prepared, one using a class A enzyme and the other a class B enzyme. The dihydronicotinamides **4.67** were converted to methanol adducts **4.68** and then ozonized to give epimeric [2-$^2$H]succinic acids **4.69** derived from atoms 3, 4, 5 and 6 of the nicotinamide ring. A reference sample of [2-$^2$H]succinic acid of known absolute stereochemistry was prepared from fumaric acid by hydration with fumarase in deuterium oxide. The stereochemistry of the product was known to be (2$S$,3$R$)-[2-$^2$H]malic acid. The hydroxyl group of the malic acid was removed by hydrogenolysis of the corresponding chloro compound to give (2$R$)-[2-$^2$H]succinic acid. Comparison of the optical rotatory dispersion curves showed that the [$^2$H]succinic acid from the class A enzymes was identical with the (2$R$)-[2-$^2$H]succinic acid from the malate whereas the class B enzyme gave a succinic acid with the opposite sign. This in turn established the stereochemistry of the dihydronicotinamide coenzymes at C-4.

The correlation of the sign of the optical rotatory dispersion curves of the stereospecifically deuteriated succinic acids with their absolute configuration enabled the stereochemistry of a number of enzymatic processes to be determined. These included the isomerization of methylmalonyl coenzyme A to succinyl coenzyme A and the conversion of (2$R$,3$S$)isocitrate to 2-oxoglutarate. The head-to-head coupling of two molecules of farnesyl diphosphate to form squalene is accompanied by the replacement of one of the hydrogen atoms from C-1 of one of the farnesyl diphosphates by a (4$S$)-hydrogen of NADPH. The stereochemistry of this process was established by degrading the squalene to give succinic acid.

The examples discussed in this chapter show how a thorough understanding of the stereochemical implications of reaction mechanisms can lead to the synthesis of labelled compounds of a well-defined stereochemistry, which in turn have been used to shed light on many biochemical processes.

## CHAPTER 5

# The Synthesis of Labelled Amino Acids

## 5.1  INTRODUCTION

α-Amino acids form the structural components of peptides and proteins. Apart from the simplest amino acid, glycine, the α-amino acids are chiral compounds. The L-enantiomer is by far the most important. Labelled amino acids have been used in metabolic studies and in determining the stereochemistry of enzyme reactions. A number of important antibiotics such as the penicillins and cephalosporins are biosynthesized from amino acids. Studies of their biosynthesis have required labelled amino acids. More recently, amino acids labelled with carbon-13 and deuterium have been used in the spectroscopic (mainly NMR) detection of interactions in protein folding.

The syntheses of labelled amino acids fall into two groups. The first group involve the synthesis of the labelled racemic DL mixtures. Many of these syntheses are based on the conventional synthetic routes to unlabelled amino acids. These are discussed in the first section. The second group of syntheses are concerned with the preparation of chirally labelled amino acids. These are discussed in the next section. A number of chemo-enzymatic syntheses use chemical methods to incorporate the label into the carbon chain of an α-keto acid and then employ an enzymatic method to convert the α-keto acid stereospecifically to the L-amino acid by transamination. A further common strategy is to use a natural amino acid as a chiral template in order to construct an enantiospecifically labelled target molecule. These methods have considerable advantages in terms of the economy of labelling. Several of the methods

The Organic Chemistry of Isotopic Labelling
By James R. Hanson
© James R. Hanson 2011
Published by the Royal Society of Chemistry, www.rsc.org

have been adapted for the preparation of nitrogen-15-labelled amino acids, which will be discussed in Chapter 7.

## 5.2 THE SYNTHESIS OF RACEMIC LABELLED AMINO ACIDS

A relatively simple procedure based on the α-bromination of a carboxylic acid and aminolysis of the bromine has been used for the preparation of labelled amino acids such as glycine from labelled acetate. The nucleophilicity of the nitrogen for this substitution reaction has been enhanced by the use of the potassium salt of phthalimide (5.1) or the potassium salt of di-*tert*-butylimide dicarbonate ($Boc_2N^-K^+$) (5.2). The imide N–H is rendered acidic by the adjacent carbonyl groups while the nitrogen anion achieves resonance stabilization by delocalization over the carbonyl groups.

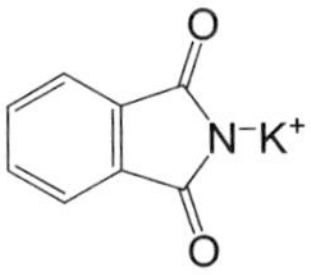

5.1 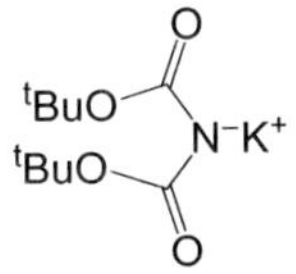          5.2

The Strecker cyanohydrin synthesis has been used on a number of occasions since the cyanide component can provide the label. An example (Scheme 5.1) comes from the synthesis of DL-[l-$^{14}$C]leucine (5.4), which was required for a study of the uptake of amino acids by liver homogenates. Isovaleraldehyde (5.3), which was obtained by degradation of inactive leucine, was treated with ammonia and potassium [$^{14}$C]cyanide. The product was then hydrolysed with sodium hydroxide to give DL-[l-$^{14}$C]leucine (5.4).

Labelled diethyl malonate has formed the starting material for the synthesis of a number of labelled amino acids. Bromination of diethyl [2-$^{14}$C]malonate and substitution of the bromine atom with the potassium salt of phthalimide give diethyl phthalimido[2-$^{14}$C]malonate (5.5).

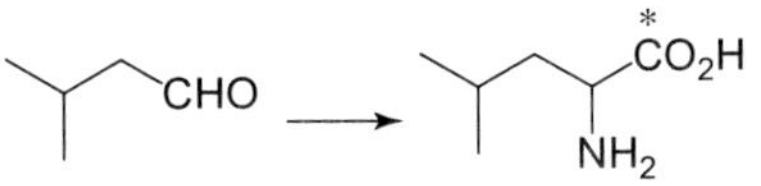

5.3          5.4

**Scheme 5.1**  The synthesis of DL-[l-$^{14}$C]leucine.

Alkylation of the remaining methine C–H of the malonate allowed the side-chain of the amino acid to be introduced. Hydrolysis of the esters and the phthalimido group followed by monodecarboxylation of the resultant dicarboxylic acid gave the DL-[2-$^{14}$C]amino acid. Ethyl acetamidocyanoacetate (**5.6**, R=CN) and diethyl acetamidomalonate (**5.6**, R=CO$_2$Et) have been used in a similar manner. These were prepared by nitrosation of ethyl cyanoacetate or diethyl malonate and reduction of the oximino tautomer of the *C*-nitroso compound with zinc and acetic acid or hydrogen over palladized charcoal in acetic anhydride. Alkylation of diethyl acetamidomalonate with ethyl 3-bromo[1-$^{14}$C]propionate, followed by hydrolysis and decarboxylation, gave DL-[5-$^{14}$C]glutamic acid (**5.7**). Alkylation of the sodium salt of ethyl [2-$^{14}$C]acetamidocyanoacetate with *N*-(δ-iodobutyl)phthalimide and hydrolysis with hydrochloric acid gave DL-[2-$^{14}$C]lysine (**5.8**). Because one of the two carboxyl groups of the malonate is lost during these syntheses, the procedure is wasteful for the preparation of carboxyl-labelled amino acids. It is useful, however, for the preparation of [2-$^{13}$C]- or [2-$^{14}$C]amino acids.

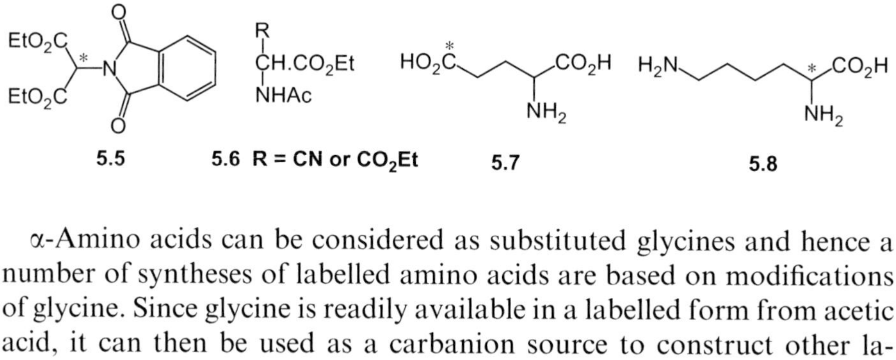

5.5              5.6  R = CN or CO₂Et        5.7                     5.8

α-Amino acids can be considered as substituted glycines and hence a number of syntheses of labelled amino acids are based on modifications of glycine. Since glycine is readily available in a labelled form from acetic acid, it can then be used as a carbanion source to construct other labelled amino acids. However, the more acidic carboxyl O–H and amine NH$_2$ need to be protected before exposing the CH$_2$ to a base to generate the carbanion. This can be achieved in several ways. In the Erlenmeyer azlactone synthesis (Scheme 5.2), glycine (**5.9**) is converted to an

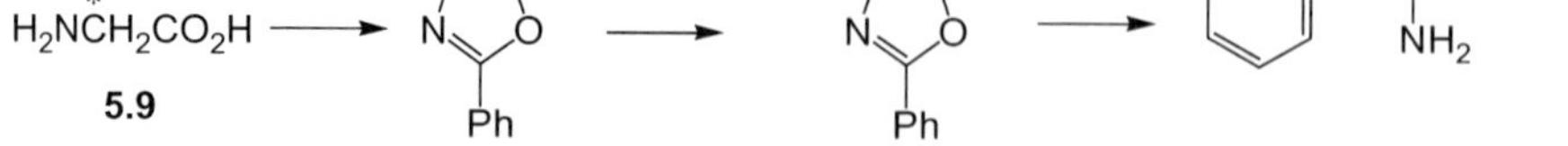

5.9              5.10              5.11              5.12

**Scheme 5.2**   The synthesis of DL-[2-$^{14}$C]phenylalanine.

azlactone (**5.10**) with benzoyl chloride *via* hippuric acid. The enol of the azlactone is part of a $6\pi$ heteroaromatic system and, with two oxygen substituents, this enol is a very electron-rich alkene. The azlactone condenses readily with aromatic aldehydes such as benzaldehyde to give the azlactone of a cinnamic acid (**5.11**). Catalytic reduction of the alkene and hydrolysis of the azlactone provide the amino acid. This method has been used to prepare DL-[2-$^{14}$C]phenylalanine (**5.12**). In the synthesis of DL-[2-$^{14}$C]valine (**5.13**), acetone was used in the condensation to give a 4-isopropylidene-2-phenyloxazolone, which was then reduced with hydrogen iodide and red phosphorus. These methods have also been used to prepare deuteriated and tritiated amino acids by catalytic reduction of the alkene using deuterium or tritium.

$$\underset{\underset{\textbf{5.13}}{}}{\overset{}{\underset{NH_2}{\overset{}{\diagup\!\!\!\!\underset{*}{\diagdown}\text{CO}_2\text{H}}}}}$$

## 5.3  LABELLED ENANTIOPURE AMINO ACIDS

In the previous section we considered syntheses of labelled racemic amino acids, many of which were carried out in the early years of the synthesis of labelled compounds. However, in recent years considerable attention has been paid to the stereospecific labelling of the enantiopure amino acids. Many of these syntheses have been achieved by combinations of chemical and enzymatic methods.

The *anti* elimination of the pro-*S* proton from C-3 by L-phenylalanine:ammonia lyase to form *trans*-cinnamic acid was established by using stereospecifically labelled L-phenylalanine. A labelled sample for these studies was prepared by converting [7-$^2$H]benzaldehyde to the azlactone derivative **5.11**. This was hydrolysed to the corresponding enamino acid. Catalytic hydrogenation gave the racemic mixture of (3*R*)-L- and (3*S*)-D-[3-$^2$H]phenylalanine. The *cis* stereochemistry of the catalytic reduction related the stereochemistry at C-2 to that at C-3. This sample was resolved by conversion to the *N*-chloroacetyl derivative. Enzymatic hydrolysis with hog kidney acylase-I liberated the free 3*R*-L isomer and left untouched the *N*-chloroacetyl-(3*S*)-D-[3-$^2$H]phenylalanine. Typical of many enzyme-catalysed reactions, phenylalanine:ammonia lyase may be used in reverse in order to prepare specifically labelled L-phenylalanines from labelled cinnamic acids. For example, L-pheny1[2-$^{13}$C,$^{15}$N]alanine has been prepared from (*E*)-[2-$^{13}$C]cinnamic acid and [$^{15}$N]ammonium chloride using phenylalanine:ammonia lyase from a yeast, *Rhodosporidium toruloides*.

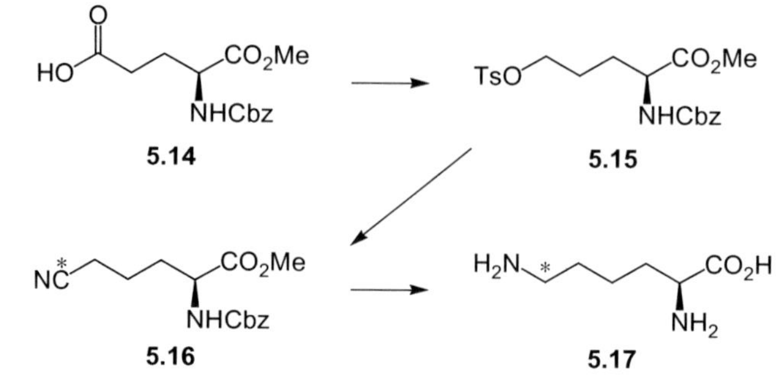

**Scheme 5.3**  The conversion of L-glutamic acid to L-[6-$^{13}$C]lysine.


The introduction of the label by modification of the side-chain of a chiral amino acid has been used, for example, in the preparation of L-[6-$^{13}$C]lysine (**5.17**), which was required for a study of the formation of collagen matrices. In this example (Scheme 5.3), the γ-carboxylic acid of *N*-benzyloxycarbonyl-L-glutamic acid α-methyl ester (**5.14**) was reduced to a primary alcohol. The label was introduced by substitution of the toluene-*p*-sulfonate of this alcohol (**5.15**) with sodium [$^{13}$C]cyanide, reduction of the nitrile **5.16** and hydrolysis of the protecting group gave L-[6-$^{13}$C]lysine (**5.17**).

The stereospecific labelling of the pro-chiral methyl groups of the L-amino acids leucine and valine is important for the study of the three-dimensional structure of proteins by NMR methods. The selective labelling of the diastereotopic methyl groups of L-leucine (Schemes 5.4 and 5.5) illustrates the role of different chiral auxiliaries in the chemo-enzymatic generation of asymmetrically labelled amino acids. The oxazolidinone **5.18**, which may be used as a chiral auxiliary, can be

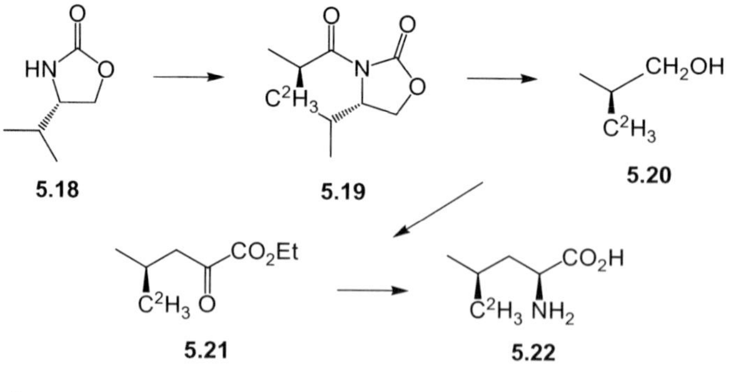

**Scheme 5.4**  An oxazolidinone as a chiral auxiliary in the selective labelling of L-leucine.

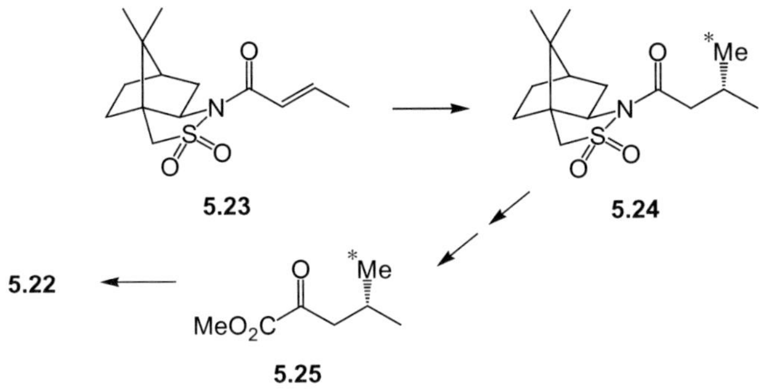

**Scheme 5.5**   A bornanesultam as a chiral auxiliary in the selective labelling of
L-leucine.

prepared from (*S*)-valinol by cyclization with phosgene and diethyl
carbonate. It readily forms *N*-acyl derivatives such as the *N*-propionyl
compound. The stereochemistry of alkylation or condensation of the
chelated (*Z*)-enolate of the amide is directed to the opposite face by the
steric hindrance of the isopropyl group of the valinol. Thus, alkylation
of the sodium enolate of the *N*-propionyl derivative with [$^2$H$_3$]methyl
iodide gave the chiral isopropyl group in **5.19**. Reductive cleavage of the
oxazolidinone released chirally labelled 2-methylpropan-1-ol (**5.20**).
This was converted to a bromo compound and thence by a Grignard
reaction with diethyl oxalate to the 2-keto ester **5.21**. Hydrolysis gave
the keto acid, which was used as a substrate for an enzymatic reductive
amination mediated by leucine dehydrogenase to form labelled L-leucine
(**5.22**).

   Another chiral auxiliary is a bornanesultam. The steric hindrance
provided by the bornyl group directs the stereochemistry of the conjugate addition to a derivative such as an *N*-crotonyl group as in **5.23**. In
this case a [$^{13}$C]dimethylcopperlithium–tributylphosphine complex was
used to introduce the label to give **5.24**. The chiral auxiliary was then
removed with lithium hydroperoxide to generate (*S*)-3-methyl[4-$^{13}$C]-
butanoic acid. A one-carbon homologation gave the α-keto ester methyl
(*S*)-4-methyl-2-oxo[5-$^{13}$C]pentanoate (**5.25**). Enzymatic hydrolysis with
*Candida rugosa* lipase and reductive amination with leucine dehydrogenase gave labelled L-leucine (**5.22**).

   The side-chains of a number of readily available α-amino acids contain functional groups which can be used in the introduction of a label,
thus avoiding the problem of resolution and the loss of labelled material.
Labelled L-aspartic acid (**5.27**) has proved to be a very useful amino acid
for the stereospecific labelling of other amino acids (Scheme 5.6). The

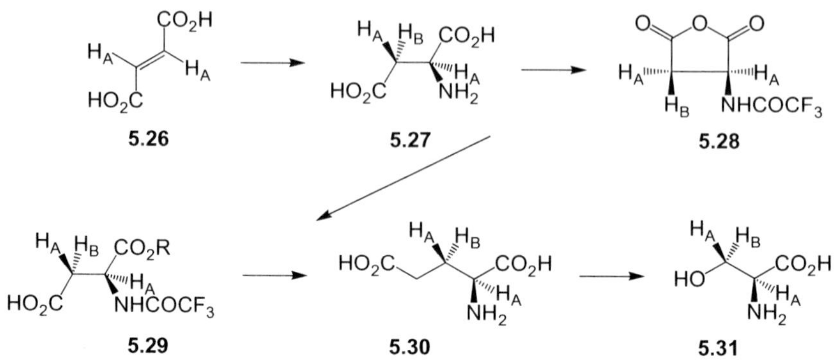

**Scheme 5.6**   The synthesis of stereospecifically labelled L-glutamic acid.

enzyme system L-aspartase will convert fumaric acid (**5.26**) to L-aspartic acid (**5.27**). The *trans* stereospecificity of this enzymatic addition of ammonia has been thoroughly established. [2,3-$^2$H$_2$]Fumaric acid (**5.26**) can be prepared by the reduction of dimethyl acetylenedicarboxylate with triphenylphosphine and deuterium oxide followed by hydrolysis. Hence by carrying out the enzymatic transformation of fumaric acid either in water or in deuterium oxide, samples of (2*S*, 3*S*)-[2, 3-$^2$H$_2$]- or (2*S*,3*R*)-[3-$^2$H$_1$]aspartic acid (**5.27**) could be prepared in useful quantities. Furthermore, the two carboxyl groups of aspartic acid can be differentiated. Treatment with trifluoroacetic anhydride gave a cyclic aspartic acid anhydride (**5.28**). Neighbouring group participation by protonation from the amide N–H favoured the regioselective ethanolysis or methanolysis of the anhydride to give largely the monoester (**5.29**, R=Me or Et). Homologation of the free carboxyl group by a Wolff rearrangement of the corresponding diazo ketone and hydrolysis afforded the stereospecifically labelled glutamic acids **5.30**. By carrying out the rearrangement in the presence of $^2$H$_2$O, a non-chiral label could be introduced at C-4. Reduction of the diazo ketone with hydrogen iodide gave a methyl ketone, which on Baeyer–Villiger oxidation with trifluoroperoxyacetic acid led to the replacement of the carboxyl group of **5.29** by a hydroxyl group and the stereospecific formation of labelled L-serine (**5.31**).

The *trans* addition of water to fumaric acid (**5.26**) by fumarase to afford (2*S*)-malic acid (**5.32**) has provided a means of preparing labelled (2*S*,3*S*)- and (2*S*,3*R*)-malates, from which some stereospecifically labelled amino acids have been obtained (Scheme 5.7). The carboxyl groups of malic acid (**5.32**) can be differentiated by reaction with para-formaldehyde in the presence of toluene-*p*-sulfonic acid, which gave a five-membered dioxolidinone (**5.33**) that protects the α-hydroxy acid.

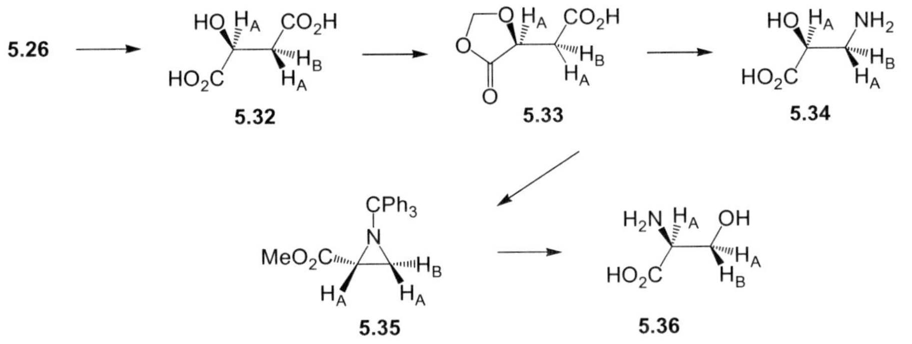

**Scheme 5.7**   The synthesis of stereospecifically labelled D-amino acids *via* isoserine.

The free carboxyl group was then replaced with an amino group by a Curtius rearrangement of the acid azide to give isoserine (**5.34**). This β-amino acid was of interest because it could be converted *via* the aziridine **5.35** to stereospecifically labelled D-amino acids $(2R,3S)$-[3-$^2$H$_1$]- and $(2R,3R)$-[2,3-$^2$H$_2$]serine (**5.36**). Other nucleophiles afforded D-cystine and D-β-chloroalanine. The isoserine derivatives have also been prepared chemically by a chiral aminohydroxylation of methyl $(Z)$- and $(E)$-[3-$^2$H$_1$]acrylates. Since the chemical method provided access to both enantiomers of the isoserine, it provided a synthesis of both L- and D-amino acids labelled at the 3-position.

The modification of the side-chain of L-glutamic acid to afford L-[6-$^{13}$C]lysine has already been described. The stereochemistry at C-2 of glutamic acid can be used to influence the stereochemistry of labelling at other centres in the molecule (Scheme 5.8). Glutamic acid undergoes cyclization to form pyroglutamic acid (**5.37**) in which the 1,3-relationship across the five-membered ring between the chiral centre at C-2 [2($S$) in the L-amino acid series] and C-4 provides the opportunity for introducing chirality at this position. Pyroglutamic acid was converted to the N-protected C-4 exomethylene derivative. Catalytic hydrogenation occurred stereospecifically to give the $(2S,4S)$-4-methyl derivative **5.38**. The steric bulk of the *tert*-butyl esters allowed a selective hydrolysis of the cyclic amide to give a 4-methylglutamic acid in which the nitrogen was protected as the Boc derivative and the α-carboxylic acid as its *tert*-butyl ester. The γ-carboxyl group was converted to a trideuteriomethyl group by reduction of a mixed anhydride with sodium borodeuteride to an alcohol and then reduction of the corresponding alkyl iodide with sodium cyanoborodeuteride. Deprotection then afforded $(2S,4R)$-[5,5,5-$^2$H$_3$]leucine (**5.39**).

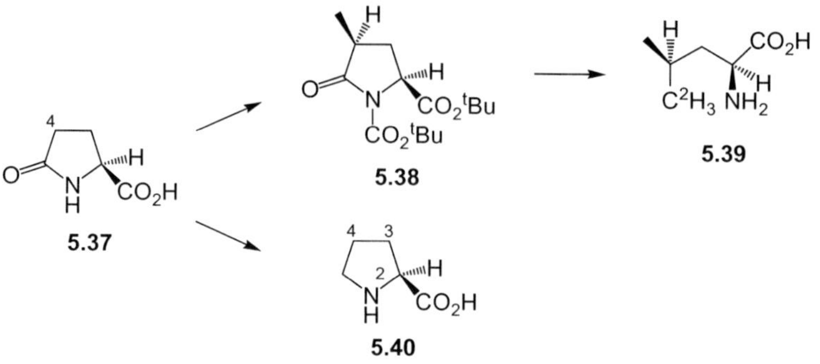

**Scheme 5.8**   Pyroglutamic acid in stereospecifically labelling amino acids.

Glutamic acid also provided a route to the labelled amino acid proline **(5.40)**. Carefully controlled conditions permitted the conversion of glutamic acid **(5.30)** stereospecifically labelled at C-2 and/or C-3 to methyl pyroglutamate. This was then converted to the protected *N*-Boc derivative. Reduction of the cyclic amide with borane–dimethyl sulfide and deprotection afforded $(2S,3S)$-[3-$^2H_1$]- and $(2S, 3R)$-[2,3-$^2H_2$]proline **(5.40)**.

The stereochemical influence of the C-2 carboxylate on labelling of the proline ring has been observed in several other situations. The stereochemistry of proline 4-hydroxylase was explored by preparing the epimeric 4-deuterioprolines. $(2S,4R)$-4-Hydroxyproline was converted to the *N*-Boc-4-toluene-*p*-sulfonate while the epimeric $(4S)$-toluene-*p*-sulfonate was obtained by reduction of the 4-ketoproline with sodium borohydride. In each case deuterium was specifically incorporated by displacement of the toluene-*p*-sulfonate with inversion of configuration using lithium triethylborodeuteride. Deuterium labels were stereospecifically introduced at C-3 by hydrolysis of the protected silylenol ether **5.41**. The addition proceeded from the *re* face. The $3(R)$ or $3(S)$ isotopomers could be selectively prepared depending on whether the hydrolysis was carried out with acidified deuterium oxide or whether a deuteriated starting material was used.

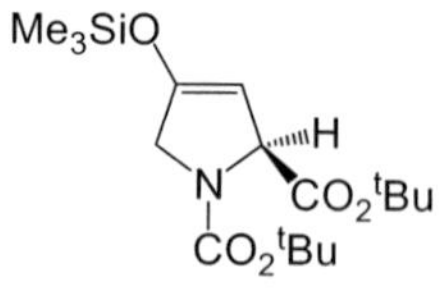

**5.41**

Chirally labelled $(2R,3R)$- and $(2R,3S)$-cysteines **(5.44)** in which there is a chiral label on the methylenethiol were prepared (Scheme 5.9) in

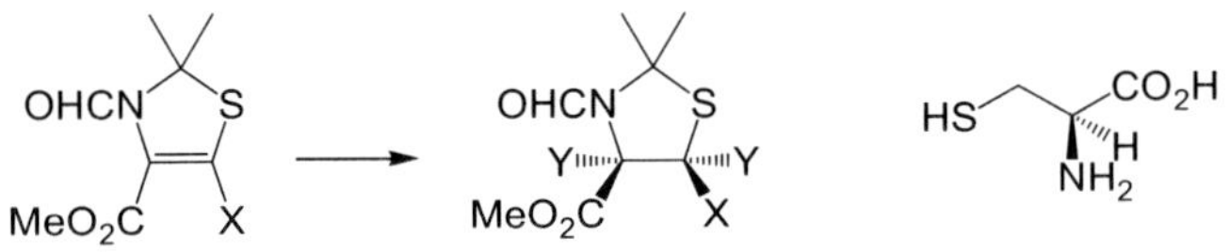

**Scheme 5.9** The synthesis of labelled cysteine.

order to study the origin of the hydrogen at C-5 in penicillin and cephalosporin biosynthesis. Pyruvic acid and [$^3$H$_4$]pyruvic acid were used as the starting materials. The strategy was to relate the stereo-chemistry of the label at C-3 to that of the amino acid at C-2 and to resolve the C-2 enantiomers. The pyruvic acid was converted to the *N*-formyl-4-thiazoline (**5.42**), which was then catalytically hydrogenated or tritiated to afford the racemic thiazolidine (**5.43**). The ester in the racemic thiazolidine was hydrolysed to the acid, which was resolved as its strychnine salt. Further hydrolysis of the heterocyclic ring gave chirally labelled samples of L-cysteine (**5.44**).

In this chapter, we have seen how the general strategies of amino acid synthesis have been adapted to the synthesis of labelled amino acids. We have also seen how chemo-enzymatic methods have evolved to give labelled materials and the way in which the chirality at one centre of an amino acid has been relayed to introduce a chiral label at another centre.

CHAPTER 6

# The Labelling of Some Compounds of Pharmaceutical Interest

## 6.1 INTRODUCTION

Many compounds that have been developed by the pharmaceutical industry have been prepared in an isotopically labelled form in order to study their pharmacokinetics, their metabolism and their binding to specific receptors or enzyme systems. The organic chemistry involved in the labelling of some examples has already been described in previous chapters, and other examples involving labelling with heteroatoms or with short-lived isotopes for positron emission tomography will be described in later chapters. In this chapter, we describe some examples of labelling strategies involving carbon and hydrogen isotopes which illustrate particular solutions to the labelling of compounds of pharmaceutical interest.

In pharmaceutical studies, there are important constraints on the isotope that is used, on the method of detection and on the site of labelling. Thus the use of carbon-14 and tritium is inappropriate in human subjects and the detection of the label may involve the separation of sub-milligram quantities of a metabolite from a highly complex mixture. The metabolic changes that affect a drug can significantly alter its chemistry. Phase 1 metabolic changes involve not only hydroxylation and dealkylation but also the hydrolysis and cleavage of a compound into smaller fragments. In order to detect all of these, it may be necessary to consider labelling a drug at a number of different sites. Phase 2

The Organic Chemistry of Isotopic Labelling
By James R. Hanson
© James R. Hanson 2011
Published by the Royal Society of Chemistry, www.rsc.org

conjugation involves the attachment of water-solubilizing groups such as glucuronic acid or sulfate esters and detoxifying units such as glutathione. These changes affect not only the site of labelling but also the experimental method that has to be used in isolating the metabolite, detecting the label and the limits of detection.

## 6.2  LIDOCAINE

A number of these points are illustrated by the labelling of the widely used local anaesthetic lidocaine (xylocaine) (**6.3**) (Scheme 6.1). This compound was introduced over 60 years ago and is used in dentistry and in minor surgery. Its synthesis involves the reaction of 2,6-dimethyl-aniline (**6.1**) with chloroacetyl chloride to form a chloroacetanilide (**6.2**). The halogen in the chloroacetyl group is then displaced by diethylamine to form lidocaine (**6.3**). This synthesis readily lends itself to labelling of the aromatic ring and the side-chain with both carbon and hydrogen isotopes. These studies have shown that the major metabolic changes involve *N*-dealkylation, hydrolysis of the amide to form *N*-diethyl- and *N*-monoethylglycine together with 2,6-dimethylaniline, oxidation of the amide to a hydroxamic acid and hydroxylation of the aromatic ring both in the 3- and 4-positions and on the methyl group. Given the various possible combinations of these processes, it is not difficult to see how the administration of this one simple compound can lead to a wide range of metabolites. Consequently, labelling just in the side-chain might lead to metabolites arising from cleavage of the 2,6-dimethylaniline portion being missed. Another more subtle point has been revealed by studies on the *N*-dealkylation. The dealkylation of $[^2H_4]$lidocaine and $[^2H_6]$lido-caine specifically deuteriated on the methylene and methyl portions of the *N*-ethyl groups was studied in rat liver microsomes. These showed primary and secondary isotope effects consistent with C–H bond breaking in the rate-determining step in the dealkylation. Similar effects have been observed in the biological demethylation of $[N\text{-}methyl\text{-}^2H_3]$-morphine. Potential isotope effects need to be taken into account when labelling a pharmaceutical substance in order to obtain a quantitative

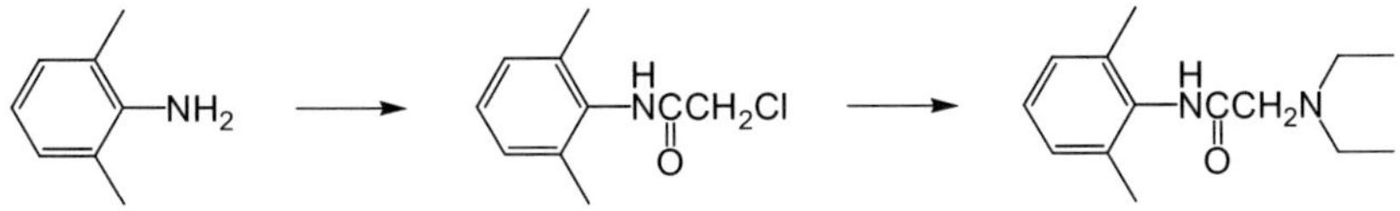

**6.1**                    **6.2**                    **6.3**

**Scheme 6.1**  The synthesis of lidocaine.

estimate of various metabolites. The labelled [$^2$H$_6$]diethylamine required for this work was prepared by the hydrogenation of [$^2$H$_3$]acetonitrile in the presence of [$^2$H$_6$]acetic anhydride followed by the reduction of the [$^2$H$_6$]-*N*-ethylacetamide with lithium aluminium hydride.

The metabolic studies on lidocaine revealed another problem, which was solved with deuteriated material. Two imidazolidinones, **6.4** and **6.5**, were detected amongst the metabolites. Using [$^2$H$_5$]ethanol and [$^2$H$_6$]lidocaine, it was shown that one imidazolinone arose from a condensation and cyclization under physiological conditions between the enzymatic *N*-dealkylation product of lidocaine and acetaldehyde formed by the biological oxidation of ethanol which was also present in the living system. The other imidazolinone arose from formaldehyde.

6.4   R = Me
6.5   R = H

## 6.3   NON-STEROIDAL ANTI-INFLAMMATORY AGENTS

Non-steroidal anti-inflammatory agents (NSAIDs) are widely used not only in general pain relief but also to reduce the inflammation and pain arising from various arthritic conditions. The profen group is exemplified by ibuprofen (Nurofen$^{\circledR}$) (**6.8**) and naproxen (**6.19**). There have been many studies of their metabolism using labelled material, the preparation of which illustrates some further strategies. The isobutyl side-chain of ibuprofen undergoes metabolic transformations involving hydroxylation of the methine and methyl carbon atoms. The latter is subsequently oxidized to a carboxylic acid. The chiral centre at C-2 in the propionic acid side-chain undergoes inversion of configuration from the *R*-enantiomer to the more biologically-active *S*-enantiomer. Naproxen undergoes demethylation to the C-6 phenol. These metabolic changes clearly influence the sites of labelling of these drugs.

The synthesis (Scheme 6.2) of ibuprofen (**6.8**) involves the Friedel–Crafts reaction of isobutylbenzene with acetyl chloride to form 4′-iso-butylacetophenone (**6.6**). The carboxyl group has then been introduced in several different ways including *via* a cyanohydrin (**6.7**) and reductive elimination of the benzylic tertiary alcohol. Another method involves a palladium-catalysed carbonylation of 1-(4-isobutylphenyl)ethanol. These syntheses offer several steps at which a label may be introduced into ibuprofen. A simple deuteriation of the α-position of ibuprofen has

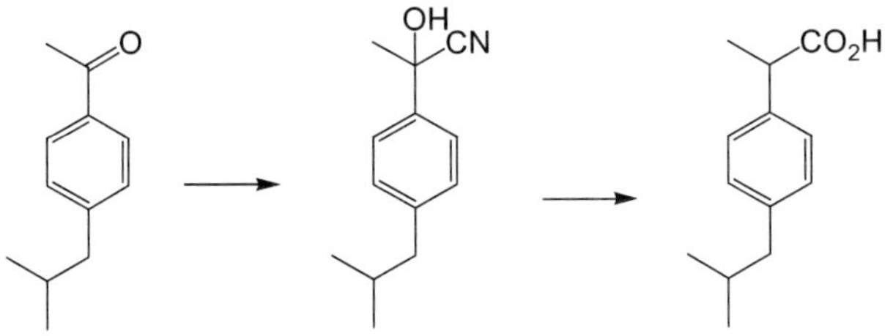

**Scheme 6.2**   The synthesis of ibuprofen.

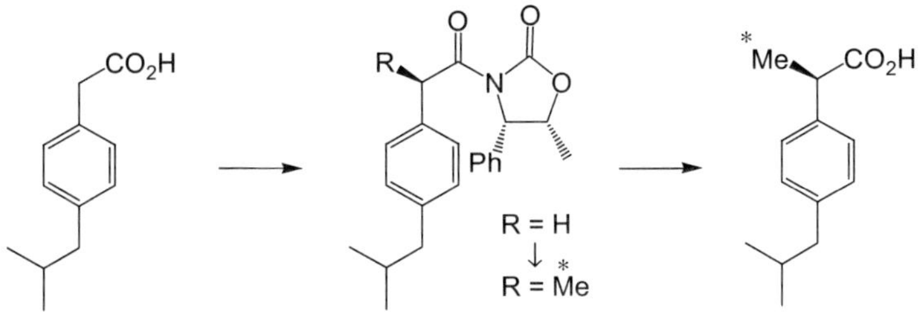

R = H
↓ *
R = Me

**Scheme 6.3**   The synthesis of (*R*)-ibuprofen.

also been achieved by an exchange reaction of the methyl ester using [$^2$H$_4$]methanol and sodium methoxide followed by careful hydrolysis.

The asymmetric synthesis (Scheme 6.3) of labelled (*R*)-ibuprofen was achieved using a chiral auxiliary, 4(*R*)-methyl-5(*S*)-phenyl-2-oxazolidi-none. *N*-Acylation of the oxazolidinone with 4′-isobutylphenylacetic acid (**6.9**) gave the derivative **6.10**. Alkylation with [$^{14}$C]methyl iodide created the chiral centre. Finally, removal of the chiral auxiliary by hydrolysis afforded (*R*)-[$^{14}$C]ibuprofen (**6.11**).

Material tritiated in the isobutyl side-chain has been prepared in several ways (Scheme 6.4). One method involved a Wadsworth–Emmons olefination reaction. *p*-Bromomethylphenylacetic acid (**6.12**) was converted to the diethylphosphonate **6.13** with triethyl phosphite and then condensed with acetone to form the unsaturated phenylacetic acid **6.14**. The chiral propionic acid side-chain was constructed by the oxazolidinone procedure described above and finally the alkene was reduced catalytically with tritium gas in the presence of Wilkinson's catalyst to give (*R*)-[$^3$H$_2$]ibuprofen (**6.15**).

Labelling one of the metabolites of ibuprofen illustrates the application of another methodology (Scheme 6.5). Diethyl methylmalonate

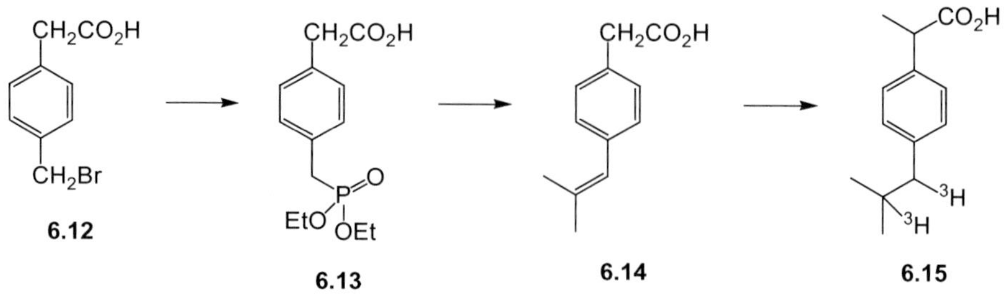

**Scheme 6.4**   The synthesis of [³H₂]ibuprofen.

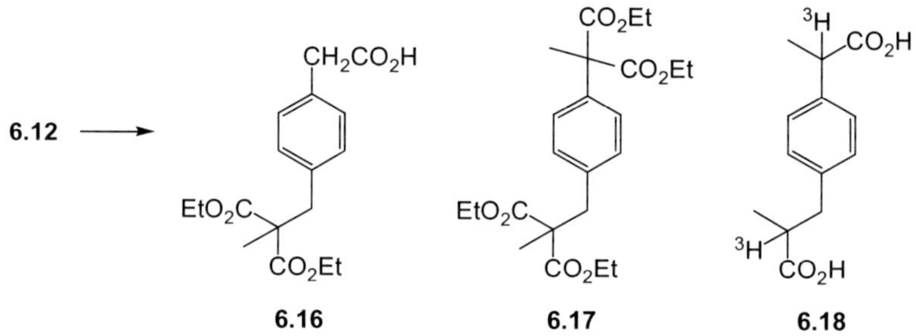

**Scheme 6.5**   The synthesis of a [³H₂]ibuprofen metabolite.

was alkylated with *p*-bromomethylphenylacetic acid (**6.12**) to give **6.16**. The triethyl ester of this was then reacted with diethylcarbonate to introduce a further ethoxycarbonyl activating group. This was alkylated with methyl iodide to give **6.17**, which contains two malonate units. Hydrolysis and decarboxylation of both of these in the presence of tritiated water gave the [³H₂]ibuprofen metabolite **6.18** labelled at both ends of the molecule.

The aromatic rings of naproxen (**6.19**) afford useful sites for labelling. The introduction of deuterium and tritium on to these rings illustrates some other methods of labelling. Treatment of the separate *R*- and *S*-enantiomers of naproxen with bromine in dichloromethane gave the 5-bromo compound **6.20**. Catalytic dehalogenation using tritium gas over palladium on charcoal gave [5-³H]naproxen (**6.21**). In another approach, the free phenol, 6-*O*-demethylnaproxen, was treated with a mixture of phosphorus pentoxide and boron trifluoride etherate in deuterium oxide to bring about exchange on the aromatic ring. In a further synthesis, alkylation of the α-position of 6-methoxynaphthyl-2-acetic acid with [$^2$H$_3$]methyl iodide gave [$^2$H$_3$]naproxen labelled in the side-chain.

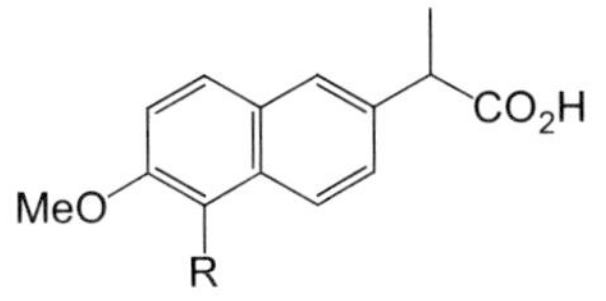

| | |
|---|---|
| **6.19** | R = H |
| **6.20** | R = Br |
| **6.21** | R = $^3$H |

## 6.4 AMPHETAMINES

Prior to achieving notoriety as drugs of abuse, amphetamines were used medicinally as central nervous system stimulants and as anorectic agents in the treatment of obesity. Labelled amphetamines have been used both in forensic analysis and in metabolic studies. The preparation of internal standards for the analysis of controlled drugs labelled with stable isotopes presents a specific problem when they are to be used in GC–MS analysis. Here it is important to introduce a sufficient number of labels into the standard to displace its molecular ion in the mass spectrum sufficiently far from that of the suspect drug sample to avoid the formation of overlapping ions which might confuse the analysis. A mass difference of at least three is recommended. In these situations, labelling a methoxy or *N*-methyl group with deuterium provides a useful strategy. As an example, [$^2$H$_6$]-2,5-dimethoxyphenylethylamine (**6.24**) has been prepared by methylating hydroquinone (**6.22**) with [$^2$H$_3$]methyl iodide (Scheme 6.6). The ethylamine side-chain was introduced by formylation with dichloromethyl methyl ether to give the aldehyde **6.23**. This was condensed with nitromethane and then the nitrostyrene was reduced with lithium aluminium hydride to afford **6.24**.

High specific activity tritiated amphetamines for metabolic studies have been prepared by the reductive tritiation of amphetamines bearing a halogen substituent on the aromatic ring. The displacement of

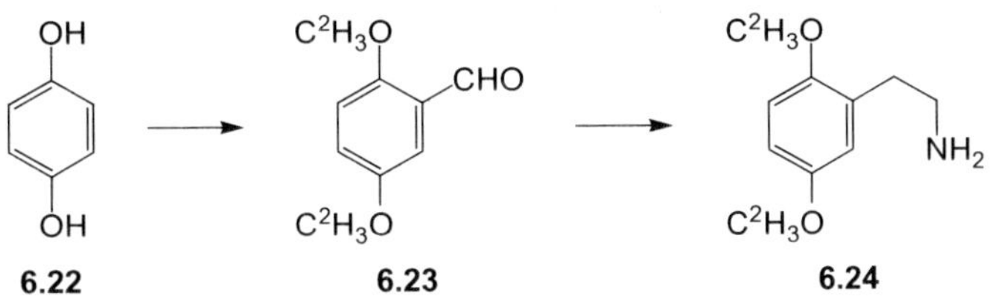

**Scheme 6.6**   The synthesis of [$^2$H$_6$]-2,5-dimethoxyphenylethylamine.

aromatic halogen substituents by tritium was carried out with tritium gas over a 10% palladium on charcoal catalyst.

## 6.5  SULFONAMIDES

Since their introduction over 70 years ago as antibacterial agents, the sulfonamides have found widespread application particularly in the treatment of urinary and respiratory infections in humans and in veterinary medicine. Apart from their action as inhibitors of folic acid biosynthesis, it has also been found that some sulfonamides are inhibitors of carbonic anhydrase and act as diuretics. The synthesis of the sulfonamide antimicrobial agents typically involves the acylation of a heterocyclic amine with 4-acetamidobenzenesulfonyl chloride (Scheme 6.7). This general synthetic scheme readily lends itself to the preparation of labelled material. Thus, [*phenyl*-U-$^{14}$C]sulfamethazine (**6.27**), which was required for metabolic studies, was prepared by the reaction of [*phenyl*-U-$^{14}$C]-4-acetamidobenzenesulfonyl chloride (**6.25**) with 2-amino-4,6-dimethylpyrimidine (**6.26**) followed by cleavage of the *N*-acetyl group. This particular sulfonamide has been widely used in veterinary medicine in, for example, the treatment of chickens, pigs and cattle. Its uptake into foodstuffs and persistence in the soil have been monitored by GC–MS using material labelled with a stable isotope as an internal standard. [*phenyl*-$^{13}$C$_6$]Sulfamethazine (**6.27**) was prepared *via* 4-acetamido[$^{13}$C$_6$]benzenesulfonyl chloride. The molecular ion of this internal standard was sufficiently distinct from that of the analyte for quantitative analysis. This and deuteriated material were also used in the identification of ions in the mass spectrum. A general method for labelling heterocyclic compounds with deuterium by exchange from deuterium oxide over palladium on charcoal has been applied both to the 2-amino-4,6-dimethylpyrimidine moiety and to sulfamethazine itself. In both cases the methyl groups were almost completely exchanged and, in the case of sulfamethazine, there was also some exchange *ortho* to the 4-amino group.

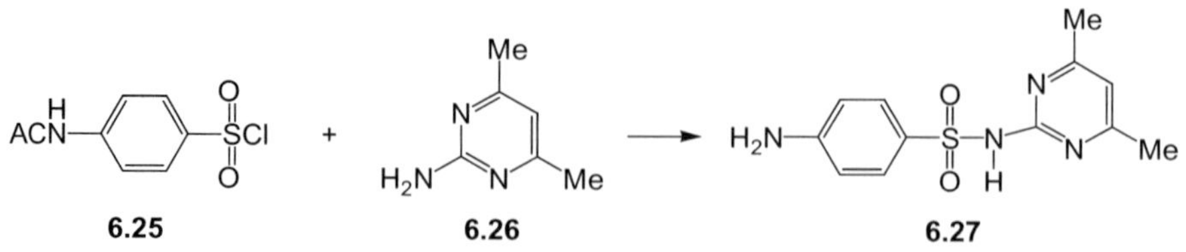

**Scheme 6.7**   The synthesis of sulfamethazine.

Sulfamethoxazole (**6.28**) is also widely prescribed for both human and agricultural use and consequently the environmental load has to be monitored. In this case, exchange from deuterium oxide containing 5% [$^2$H$_2$]sulfuric acid led to the incorporation of the label at the C-3 and C-5 positions *ortho* to the aromatic amine and at C-4′ on the oxazole ring.

**6.28**

## 6.6  MORPHINE AND ITS RELATIVES

Morphine (**6.29**) and some of its relatives are powerful painkillers, although they are also addictive narcotic agents. There are specific receptors in the brain known as the opioid receptors which respond to morphine and its analogues. There has been considerable effort to synthesize derivatives which possess enhanced painkilling properties but are non-addictive. Consequently, there have been studies not only on the preparation of labelled samples of morphine but also on its phenolic methyl ether codeine (**6.30**) and their relatives for both metabolic and forensic purposes. Morphine is a chiral natural product and its total synthesis has required substantial effort. Unlike the compounds discussed so far in this chapter, economic considerations therefore dictate that the natural product itself must form the starting material for labelling.

| 6.29  R = H | 6.31  R = H | 6.33 |
| 6.30  R = Me | 6.32  R = SiMe$_2$$^t$Bu | |

Generally labelled [$^3$H]morphine has been prepared by the Wilzbach process. [*N-methyl*-$^{14}$C]Morphine was prepared from normorphine by reductive methylation with [$^{14}$C]paraformaldehyde in ethanol containing some formic acid. In this example, reductive methylation avoids the potential problem of over-methylation and the formation of quaternary ammonium salts. This product has been converted to the biologically active C-6 glucuronide. However, from the point of view of metabolic studies, *N*-demethylation is a significant biotransformation of morphine. Consequently, other sites have been labelled. Thus, [6-³H]morphine has been prepared by reduction of the 6-ketone, morphinone, with sodium [³H]borohydride. Morphine and codeine have been prepared labelled at both C-1 and C-2 on the aromatic ring. Iodination of codeine with chloramine-T and sodium iodide in 0.1 M hydrochloric acid gave 1-iodocodeine (**6.31**) in a procedure which has also been used to prepare [¹²⁵I]iodocodeine. The methoxyl group of the iodocodeine was cleaved with boron tribromide to give the phenol 1-iodomorphine. Conversion of the C-6 allylic alcohols of the 1-iodomorphine and 1-iodocodeine to their bulky *tert*-butyldimethylsilyl ethers (**6.32**) served to protect the adjacent 7(8)-double bond, allowing catalytic tritiolysis of the iodine atoms to be carried out to give [1-³H]morphine (**6.29**) and [1-³H]codeine (**6.30**).

The preparation of carbon-labelled benzylisoquinoline intermediates for the study of the biosynthesis of morphine was described in Chapter 2. However, the sequence which inter-related codeinone (**6.33**), codeine (**6.30**) and morphine (**6.29**) in the opium poppy was studied using substrates labelled with tritium at C-2. [2-³H]Morphine (**6.29**) was prepared by the base-catalysed exchange of the proton adjacent to the phenol using tritiated water in dimethylformamide. The labelled product was methylated with trimethylanilinium hydroxide to form [2-³H]codeine (**6.30**). Mild oxidation with silver carbonate gave [2-³H]codeinone (**6.33**), from which [2,6-³H₂]codeine was obtained by reduction with sodium [³H]borohydride.

[*O-methyl*-¹⁴C]Codeine (**6.30**) has been prepared from morphine. Simple methylation of morphine with [¹⁴C]methyl iodide had the possibility of also leading to quaternization of the tertiary nitrogen. Consequently, the nucleophilicity of the nitrogen of morphine was reduced by converting it to the *N*-oxide. The phenolic hydroxyl in this protected derivative was then methylated with [¹⁴C]methyl iodide to give [*O-methyl*-¹⁴C]codeine *N*-oxide. The *N*-oxide was then reduced with sulfur dioxide to form [*O-methyl*-¹⁴C]codeine.

One of the morphine derivatives which has been labelled (Scheme 6.8) is the analgesic etorphine (**6.38**). Deuterium-labelled [6-O-*methyl*-²H₃]-codeine was prepared from codeine by methylation with [²H₃]methyl iodide. Dehydrogenation with DDQ gave the diene thebaine (**6.34**). The ethano bridge of etorphine was constructed by a Diels–Alder reaction with acrylonitrile. Reaction of the resultant nitrile **6.35** with

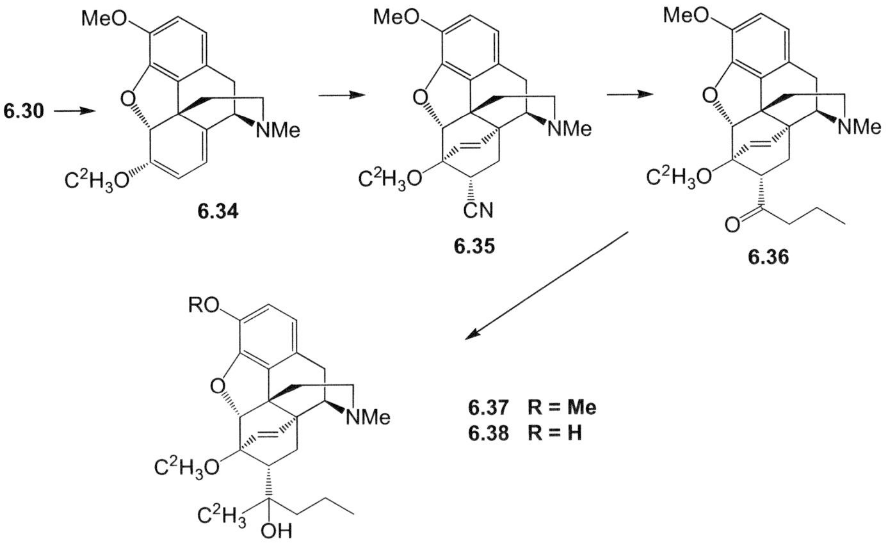

**Scheme 6.8**   The synthesis of [$^2$H$_6$]etorphine.

*n*-propylmagnesium bromide gave the butanone **6.36**. Another Grignard reaction with [$^2$H$_3$]methylmagnesium iodide was used to introduce a second trideuteriomethyl group to give **6.37**. Cleavage of the phenolic methoxyl group then gave labelled etorphine (**6.38**).

The aporphines, which are derived from morphine and codeine, exhibit interesting biological properties as rigid dopamine agonists. This rigidity imparts a selectivity in terms of the receptors to which they bind. They have application in the treatment of Parkinson's disease. The ring system, which is simpler than that of morphine and has only one chiral centre, is amenable to total synthesis (Scheme 6.9). Thus [6α-$^{14}$C]apomorphine (**6.44**) has been synthesized by two cyclizations of the amide **6.41**, which was obtained from 3,4-dimethoxy-2-nitrophenyl[*carbox-yl*-$^{14}$C]acetyl chloride (**6.40**) and β-phenylethylamine. The carbon-14 label was introduced by treatment of 3,4-dimethoxy-2-nitrobenzyl chloride (**6.39**) with potassium [$^{14}$C]cyanide. Hydrolysis of the nitrile and treatment with oxalyl chloride gave the acid chloride **6.40**, which was required for the preparation of the amide **6.41**. Application of a Bischler–Napieralski cyclization to the amide **6.41** using phosphorus pentoxide and phosphorus oxychloride led to the dihydroisoquinoline **6.42**. Reduction of the imine with sodium borohydride and methylation of the isoquinoline amine was followed by reduction of the nitro group with zinc dust and hydrochloric acid to give another amine, **6.43**. A copper-catalysed Pschorr cyclization of the diazonium salt derived from

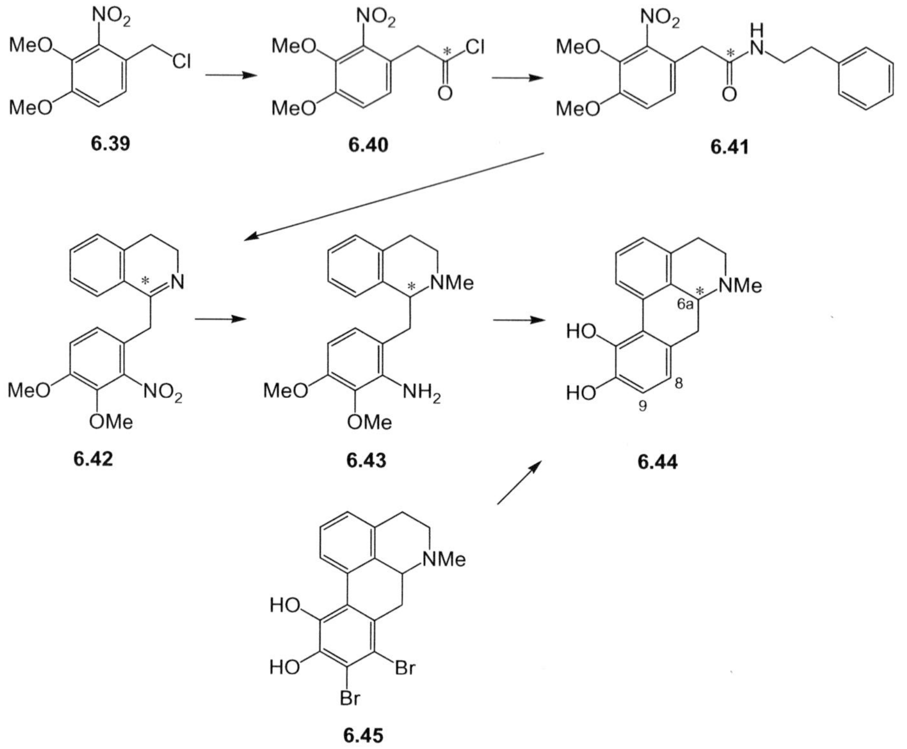

**Scheme 6.9**   The synthesis of labelled apomorphine.

this amine afforded [6α-¹⁴C]apomorphine dimethyl ether. The racemate was resolved and demethylated to give the labelled apomorphine (**6.44**).

Tritiated material was prepared more easily. Bromination of apomorphine (**6.44**) in trifluoroacetic acid gave 8,9-dibromoapomorphine (**6.45**). Reduction with tritium over 10% palladium on charcoal in ethanol yielded (−)-[8,9-³H₂]apomorphine. Comparison of these two syntheses shows the importance of selecting the isotope and site of labelling when considering the economics of a strategy. Not only is the first sequence considerably longer than the second, with the label being introduced early in the synthesis, but also half of the labelled material is lost in the resolution stage, which occurs after the label has been introduced.

## 6.7   GALANTHAMINE

The alkaloid galanthamine (**6.46**), which is obtained from several species of the Amaryllidaceae, is used in the treatment of Alzheimer's disease.

The structural complexity of the alkaloid rules out total synthesis as an efficient labelling strategy. However, like many alkaloids, galanthamine possesses both *O*-methyl and *N*-methyl groups, which for some but not all purposes provide suitable sites for introducing a label. [6-*O*-*methyl*-$^2$H$_3$,*N*-*methyl*-$^2$H$_3$]Galanthamine (**6.46**) has been obtained from galanthamine by selective *O*- and *N*-demethylation followed by the re-introduction of labelled methyl groups. This sequence reveals the need to develop methods for the selective degradation of natural products in the context of preparing labelled material. The methoxyl group of galanthamine was hydrolysed by the action of lithium Selectride$^\circledR$ and the *N*-methyl group was removed by the action of iron salts on the *N*-oxide to give a bisnorgalanthamine (**6.47**). The secondary amine of the bisnorgalanthamine was protected as its *tert*-butyloxycarbonyl deriva-tive (**6.48**), allowing the phenolic hydroxyl group to be methylated with [$^2$H$_6$]dimethyl sulfate to give **6.49**. The Boc protecting group was re-moved and, in order to avoid the formation of a quaternary ammonium salt, the amine was reductively methylated with [$^2$H$_2$]formaldehyde, [$^2$H]acetic acid and sodium [$^2$H$_4$]borohydride to afford the labelled galanthamine (**6.46**).

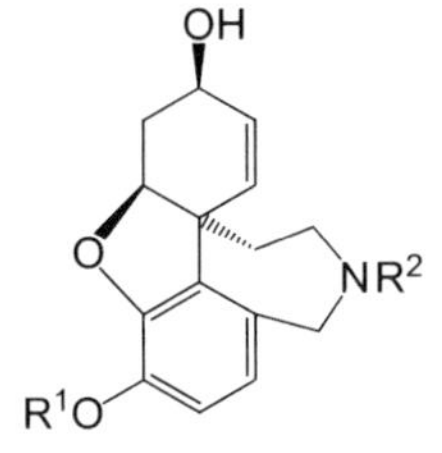

6.46   R$^1$ = R$^2$ = Me
6.47   R$^1$ = R$^2$ = H
6.48   R$^1$ = H, R$^2$ = Boc
6.49   R$^1$ = Me, R$^2$ = Boc

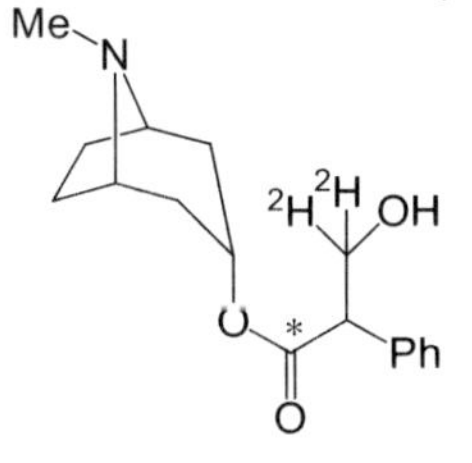

6.50

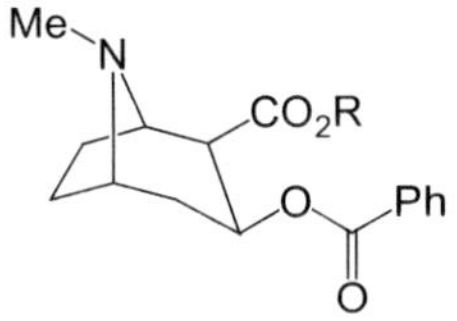

6.51   R = Me
6.52   R = Et

## 6.8   TROPANE ALKALOIDS

The bicyclic tropane alkaloids include atropine (**6.50**) and cocaine (**6.51**). They have been prepared in a labelled form for biosynthetic studies and in connection with their biological activity. In the case of cocaine, deuteriated samples have been used in forensic analysis. Since cocaine binds selectively to specific centres in the brain, it has also been labelled with short half-life isotopes, carbon-11 and fluorine-18, for use in positron emission tomography. This work is described in Chapter 9.

Atropine [(±)-hyoscyamine] (**6.50**) comprises two portions, the bicyclic tropane unit and an esterifying tropic acid moiety. In cocaine there is a benzoyl ester and a carbomethoxy group attached to the tropane unit. Although the tropane unit has been prepared in labelled forms by total syntheses, these are multi-step processes. Most of the labelled samples have been prepared starting from the chiral natural product. The *N*-methyl group of atropine (**6.50**) was removed by reaction with α-chloroethyl chloroformate whereas that of cocaine (**6.51**) was removed with vinyl chloroformate. In both cases the nortropanes were then remethylated with [$^{14}$C]methyl iodide.

Cocaine is rapidly metabolized and at least 11 metabolites have been detected in human urine. These include ecgonine, which arises by hydrolysis of both esters, the partial hydrolysis products, benzoylecgonine and ecgonine methyl ester and hydroxylation products such as *m*-hydroxybenzoylecgonine. When cocaine is ingested with alcohol, *trans*-esterification also takes place, possibly facilitated by neighbouring group participation of a protonated quaternary salt of the nitrogen bridge. This gives the ethyl ester, cocaethylene (**6.52**). This metabolite has been found in substantial amounts in the biological fluids and tissues of people who have died after using both cocaine and ethanol for 'recreational' purposes. In order to quantify this metabolite and the parent cocaine by GC–MS, deuteriated standards have been prepared from norcocaine by *N*-methylation with [$^2$H$_3$]methyl iodide and by preparing the trideuteriomethyl and trideuterioethyl esters of benzoylecgonine.

The $^{13}$C,$^2$H doubly labelled tropic acid portion of the alkaloid hyoscyamine (**6.50**), which was required for biosynthetic studies, has been prepared by the base-catalysed condensation of [1'-$^{13}$C]phenylacetyltropine with [$^2$H$_2$]formaldehyde.

## 6.9   TRYPTAMINE DERIVATIVES

There has been considerable interest in the neurochemical activity of tryptamine derivatives such as serotonin (5-hydroxytryptamine) and melatonin (*N*-acetyl-5-methoxytryptamine) and in drugs of abuse such as psilocybin (the monophosphate of *N,N*-dimethyl-4-hydroxytryptamine). Labelled samples have been required for metabolic studies and as standards for forensic use. There are a number of syntheses (Scheme 6.10) based on the acylation with oxalyl chloride of an appropriately substituted indole followed by reaction with an amine to

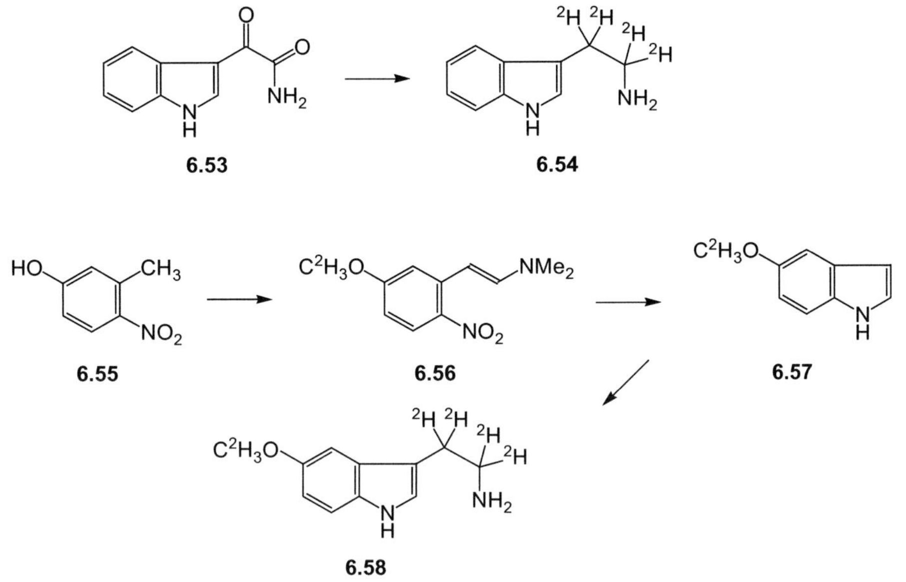

**Scheme 6.10** The synthesis of labelled tryptamine derivatives.

give an indol-3-ylglyoxalylamide (**6.53**). Reduction of this with lithium aluminium [$^2H_4$]hydride provided a straightforward synthesis of the $\alpha,\alpha,\beta,\beta$-$^2H_4$ derivative **6.54**.

A synthetic scheme which has allowed the introduction of labels at various centres in tryptamine derivatives started with 5-hydroxy-2-nitrotoluene (**6.55**) (Scheme 6.10). The nitro group played several roles. By rendering the phenolic hydroxyl group weakly acidic, the latter was easily methylated with [$^2H_3$]methyl iodide. The nitro group also facilitated condensation of the adjacent methyl group with *N,N*-dimethylformamide dimethyl acetal to give the styryldimethylamine **6.56**. which on reductive cyclization with hydrogen over palladium on charcoal formed [5-*O-methyl*-$^2H_3$]indole (**6.57**). The side-chain was then added by reaction with oxalyl chloride and dimethylamine followed by reduction with lithium aluminium [$^2H_4$]hydride to afford **6.58**.

In this chapter, we have seen examples of several different labelling strategies, some based on total synthesis and others on modifying a natural product. The selection of the isotope, the site of labelling relative to possible metabolic pathways and the application of the labelled product are all features which have to be taken into account in determining the strategy to be used.

CHAPTER 7

# Labelling Compounds with the Stable Isotopes of Nitrogen and Oxygen

In this chapter we consider the methods that have been used to introduce the stable isotopes of nitrogen and oxygen into organic compounds. These isotopes have played an important role both in the study of organic reaction mechanisms and in biosynthesis.

## 7.1 NITROGEN-15

There are two isotopes of nitrogen, $^{13}N$ and $^{15}N$, that are used in labelling experiments. Nitrogen-13 is a positron-emitting isotope with a short half-life which is used in positron emission tomography. This application is discussed in Chapter 9. Nitrogen-15 is a stable isotope which is present in 0.38% natural abundance. Although it can be separated from nitrogen-14 by the fractional distillation of ammonia or nitric oxide, the common method of separation involves chemical exchange between the oxides of nitrogen and nitric acid. An older method used the exchange between gaseous ammonia and aqueous ammonium salts. The isotope effects in these exchange reactions mean that it is possible to produce enrichments of up to 99.9 atom% of nitrogen-15. The processes are, however, highly corrosive. The enriched nitric acid can be converted to nitric(III) (nitrous) acid and metal nitrates and reduced to ammonia.

The Organic Chemistry of Isotopic Labelling
By James R. Hanson
© James R. Hanson 2011
Published by the Royal Society of Chemistry, www.rsc.org

This stable isotope of nitrogen can be detected by mass spectrometric and nuclear magnetic resonance methods. Signals in the $^{15}$N NMR spectrum cover a wide range of chemical shifts and these are particularly sensitive to structural and environmental changes. Variations in the chemical shift as a consequence of protonation, hydrogen bonding, metal complex formation and other interactions are particularly helpful in establishing a three-dimensional picture of protein and nucleic acid structures. Hence the labelling of compounds with nitrogen-15 is of value not just in establishing metabolic pathways and revealing mechanisms but also in determining the structures of biological macromolecules. Labelling with nitrogen-15 has also been used in unravelling the mass spectrometric fragmentation pattern of natural molecules such as the nucleic acid bases.

There are a range of primary sources of nitrogen-15 including [$^{15}$N]ammonia, sodium [$^{15}$N]nitrate, sodium [$^{15}$N]nitrite, [$^{15}$N]urea, potassium [$^{15}$N]cyanide and [$^{15}$N]phthalimide. Hence many of the conventional syntheses of organonitrogen chemistry can be repeated with nitrogen-15. Several methods described in Chapter 5 concerning the synthesis of labelled amino acids have been adapted for the introduction of a nitrogen label. The main emphasis has been to develop specific high-yielding syntheses in which the isotopic label is introduced late in a sequence.

The convenient heterocyclic building block [$^{15}$N]malonitrile can be prepared from diethyl malonate by reaction with [$^{15}$N]ammonia and dehydration of the resulting malonamide. Another useful compound for introducing a nitrogen-15 label is [$^{15}$N]benzylamine, which can be employed in nucleophilic substitutions. It is formed by the reduction of [$^{15}$N]benzonitrile or [$^{15}$N]benzamide. Once it has been used in a nucleophilic substitution reaction, the benzyl group can be removed by hydrogenolysis, leaving the nitrogen label attached to the other component. The use of benzylamine rather than ammonia reduces the chance of multiple substitution on nitrogen. Another useful reagent for nucleophilic substitution is the potassium salt of [$^{15}$N]phthalimide. Some illustrations of the use of these sources of a nitrogen-15 label are provided in the following syntheses.

The amino acid [$^{15}$N]glycine has been prepared by the action of potassium [$^{15}$N]phthalimide on ethyl bromoacetate followed by hydrolysis. The nitrogen-15-labelled glycine was then used as a source of other nitrogen-15-labelled amino acids by a hydantoin synthesis. This is exemplified by the synthesis of DL-[α-$^{15}$N]tryptophan (**7.3**) (Scheme 7.1). The ethyl ester of [$^{15}$N]glycine was converted to a urea by the action of potassium cyanate and thence by cyclization with hydrogen chloride to

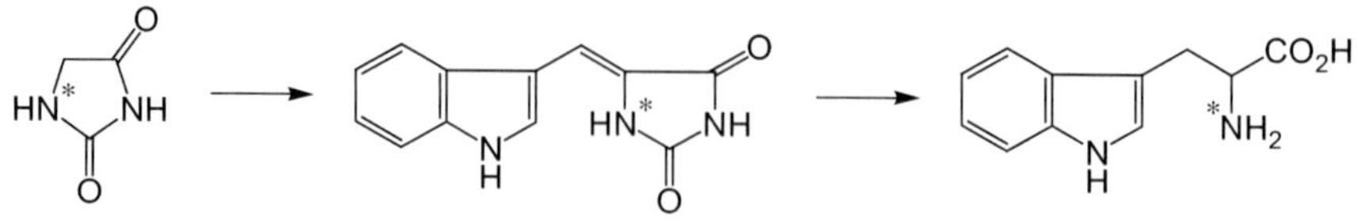

**7.1**                                    **7.2**                                    **7.3**

**Scheme 7.1**   The synthesis of DL-[α-$^{15}$N]tryptophan.

**7.4**                                                    **7.5**

**Scheme 7.2**   The synthesis of L-[α-$^{15}$N]lysine.

hydantoin (**7.1**), which is specifically labelled on only one of the two nitrogen atoms. Condensation with indolyl-3-carboxaldehyde followed by reduction of the product **7.2** and hydrolysis afforded DL-[α-$^{15}$N]tryptophan (**7.3**).

The preparation of L-[α-$^{15}$N]lysine (**7.5**) illustrates the introduction of a label at a late step and the importance of considering the stereo-chemistry of reactions (Scheme 7.2). ε-Benzoyl-D-lysine from the enan-tiomeric amino acid was treated with nitrosyl bromide to form chiral 6-benzamido-2-bromohexanoic acid (**7.4**) with retention of configuration at the α-position. The stereochemical outcome of the diazotization and bromination sequence is affected by α-lactone formation as the diazo-nium group is displaced and hence the incoming nucleophile approaches the α-carbon from the opposite face of the lactone to give the overall retention of configuration. However, the nucleophilic displacement of the bromine with [$^{15}$N]ammonia takes place with inversion of con-figuration to give, after hydrolysis of the benzoyl group, the L-amino acid **7.5**.

[$^{15}$N]Phthalimide has been used to prepare the amino acid ornithine (**7.9**) labelled on either of the two nitrogens (Scheme 7.3). The label on the δ-amino group was introduced by treating potassium [$^{15}$N]phthali-mide with 1,3-dibromopropane to give 3-bromopropyl[$^{15}$N]phthalimide (**7.6**). The second bromine was displaced by the carbanion derived from phthalimidomalonic acid diethyl ester (**7.7**), itself prepared from bro-momalonic acid diethyl ester and phthalimide. The resulting α-phthali-mido-γ-[$^{15}$N]phthalimidopropylmalonic acid diethyl ester (**7.8**) was hydrolysed and decarboxylated by boiling in a mixture of acetic and

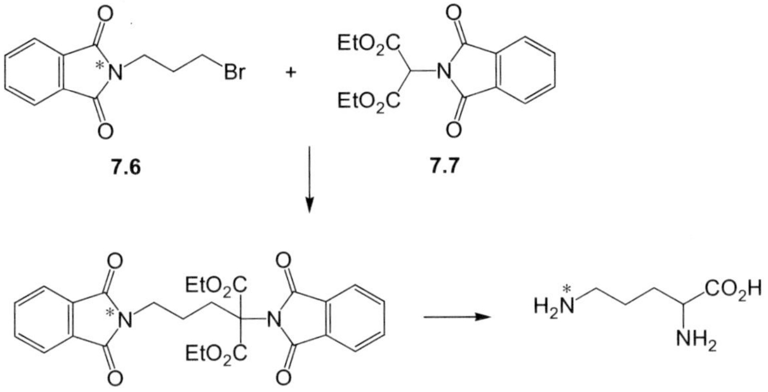

**Scheme 7.3**   The synthesis of DL-[5-$^{15}$N]ornithine.

hydrochloric acids to give DL-[$^{15}$N]ornithine (**7.9**) labelled in the δ-amino group. Labelling the α-amino group was achieved by using [$^{15}$N]phthalimidomalonic acid diethyl ester, a method which has applications in the synthesis of other nitrogen-15-labelled amino acids.

The preparation of nitrogen-15-labelled nucleic acid bases has been used in the determination of nucleic acid structures and in the identification of particular interactions. Variously labelled samples of pyrimidines and purines have been prepared by the displacement of chlorine from the chloropyrimidines and chloropurines with [$^{15}$N]benzylamine. For example, the synthetic cytokinin [$^{15}$N]-$N^6$-benzyladenine (**7.11**) was obtained using the reaction between [$^{15}$N]benzylamine and 6-chloropurine (**7.10**).

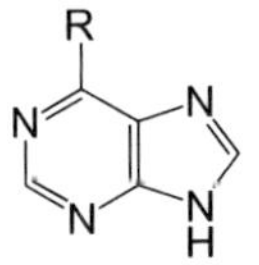

7.10   R = Cl
7.11   R = HNCH$_2$C$_6$H$_5$

Nitrosation and nitration reactions are other methods that have been used to introduce a nitrogen-15 label. These are exemplified by the syntheses of nitrogen-15-labelled nucleic acid bases (Scheme 7.4). [7-$^{15}$N]Adenine (**7.14**) was obtained by nitrosation of 4,6-diaminopyrimidine (**7.12**) with sodium [$^{15}$N]nitrite and reduction of the nitroso compounds to give the [5-$^{15}$N]triaminopyrimidine (**7.13**). The imidazole

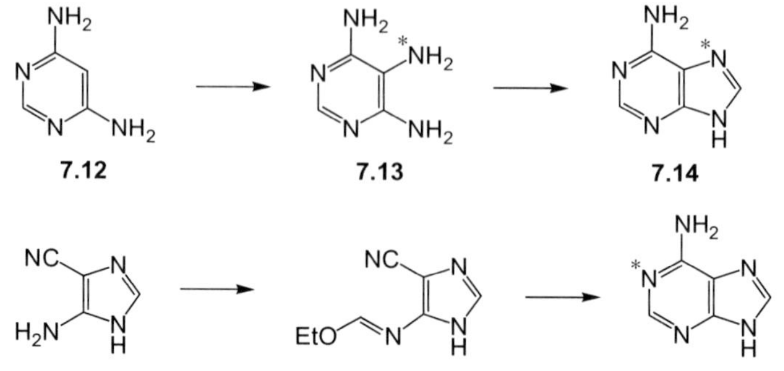

**Scheme 7.4**   The synthesis of [$^{15}$N]nucleic acid bases.

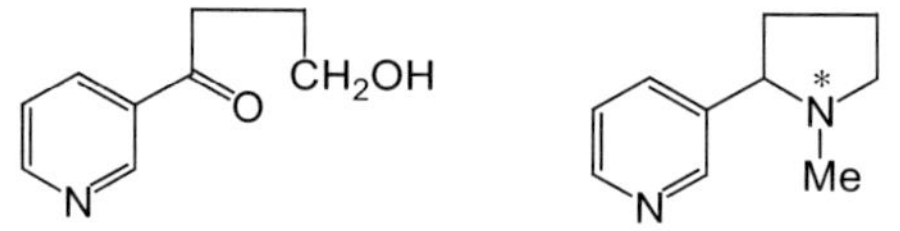

**Scheme 7.5**   The synthesis of [$^{15}$N]nicotine.

ring of adenine was created by condensation of the 5,6-diamine with a formate equivalent, diethoxymethyl acetate (acetoxydiethoxymethane). Nitrogen-15 labels were introduced into the pyrimidine ring by attaching the nitrogen labels to the preformed imidazole ring. Thus, [1-$^{15}$N]adenine (**7.17**) was obtained from 4-amino-5-cyanoimidazole (**7.15**) by first treatment with diethoxymethyl acetate to form 5-cyano-4-(ethoxymethylene)aminoimidazole (**7.16**) and then cyclization with [$^{15}$N]ammonia to give **7.17**.

    *N*-Nitrosonornicotine is a known carcinogen which is formed from nicotine in tobacco. The preparation of nicotine labelled on the pyrrolidine ring in order to study its metabolism illustrates the use of reductive amination to introduce nitrogen-15 (Scheme 7.5). Treatment of 3-bromopyridine with butyllithium and γ-butyrolactone created the carbon skeleton **7.18** of nornicotine. Mild oxidation converted the primary alcohol to an aldehyde. The pyrrolidine ring was then formed by treatment with [$^{15}$N]ammonia and sodium cyanoborohydride. *N*-Methylation gave [$^{15}$N]nicotine (**7.19**).

    Nitrogen-15 was used as a tracer in a study of the mechanism of the Fischer indole synthesis (Scheme 7.6). Benzoyl chloride was converted to [$^{15}$N]benzamide (**7.20**) by reaction with [$^{15}$N]ammonia and then

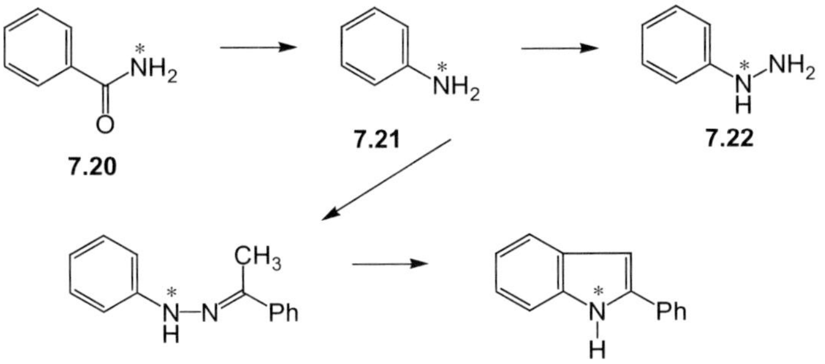

**Scheme 7.6**     $^{15}$N labelling in the Fischer indole synthesis.

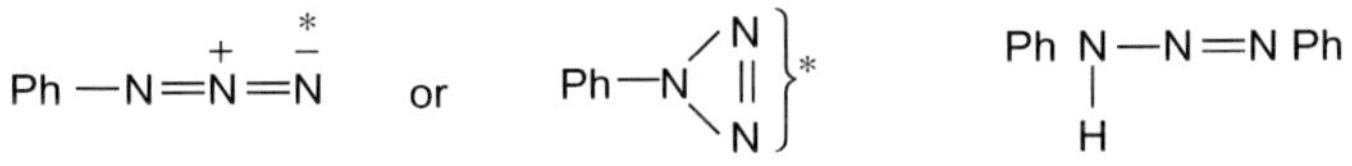

**Scheme 7.7**     The structure of phenyl azide.

degraded to [$^{15}$N]aniline (**7.21**) by a Hofmann degradation with bromine in sodium hydroxide. On diazotization and reduction this was transformed into [$^{15}$N]phenylhydrazine (**7.22**) in which the single nitrogen-15 label is attached to the aromatic ring. Acetophenone phenylhydrazone (**7.23**) was then prepared and subjected to the Fischer indole synthesis. The resulting 2-phenylindole (**7.24**) retained the nitrogen-15 label in accordance with the proposed mechanism for the reaction.

A nitrogen-15 label was used to distinguish between the linear ketene-like **7.25** and the three-membered ring formulations **7.26** for phenyl azide (Scheme 7.7). Phenyl [$^{15}$N]azide was prepared from phenylhydrazine and [$^{15}$N]nitrous acid. This was treated with phenylmagnesium bromide to give phenyl [$^{15}$N]diazoaminobenzene (**7.27**). If the linear formulation was correct, the nitrogen-15 label would be restricted to the terminal position of **7.27** whereas the cyclic formulation would place some of the nitrogen-15 label on the central nitrogen atom of this Grignard adduct. The distribution of the nitrogen-15 label in the products of reductive cleavage of the phenyldiazoaminobenzene showed that the label was restricted to the terminal position of the N$_3$ unit in accord with the linear formulation.

Chiral amino acids that are labelled with nitrogen-15 have been prepared by the enzymatic reductive amination of a range of α-keto acids.

Thus, L-[3-$^{13}$C,$^{15}$N]alanine has been prepared from [3-$^{13}$C]pyruvic acid and [$^{15}$N]ammonium formate in the presence of alanine dehydrogenase. Leucine dehydrogenase and phenylalanine dehydrogenase were used in a similar way to introduce nitrogen-15 labels into the respective amino acids starting from the corresponding α-keto acids.

## 7.2   OXYGEN-17 AND OXYGEN-18

Oxygen-17 and oxygen-18 (0.02% and 0.2% natural abundance, respectively) are stable isotopes of oxygen. The use of the radioactive isotope oxygen-15 for medical imaging is discussed in Chapter 9. The stable isotopes can be separated by the fractionation of water particularly as part of the processes that are directed at the purification of deuterium. The most readily available isotope is oxygen-18. For many studies, an enrichment of up to 25% is all that is required. Oxygen-18 of this enrichment can also be obtained by the distillation of oxygen, although this is limited by the fact that oxygen-18 is present in dioxygen mainly as $^{16}$O–$^{18}$O. When the isotope distribution is considered from experiments using 'oxygen-18' gas, it is important to consider whether the gas is $^{18}$O–$^{18}$O perhaps containing some $^{16}$O–$^{16}$O or whether it is just $^{16}$O–$^{18}$O. Some important experiments such as that concerning the role of an endoperoxide in the origin of the oxygen atoms at C-9 and C-11 of the prostaglandins relied on the use of $^{18}$O–$^{18}$O oxygen and showing by mass spectrometry that two atoms bearing oxygen-18 were incorporated into the same molecule. Increasing amounts of highly enriched [$^{18}$O]water are being made for use in the preparation of fluorine-18 for positron emission tomography. The presence of oxygen-18 in a molecule may be established by mass spectrometry or by the isotope shift on an attached carbon-13 NMR signal whereas oxygen-17 (spin 5/2) can be observed directly in the NMR spectrum. In some biosynthetic experiments, a $^{13}$C,$^{18}$O doubly labelled substrate has been used. The $^{13}$C NMR signal of the carbon-13 label acts as a 'reporter' for the presence of the oxygen-18 label.

Some of the earliest experiments with oxygen-18 provided a basis for understanding the mechanism of ester hydrolysis and are described in Chapter 1.

Many biological processes involve the insertion of oxygen. The origin and the fate of oxygen atoms in compounds of biological importance have been studied using oxygen labelling. Although isotope fractionation arising from isotope effects which can subtly affect the oxygen-18:oxygen-16 ratio has been used in archaeological, geochemical and some biochemical and analytical studies, the majority of experiments have been carried out with enriched material in which isotope effects have been of less concern.

The stability of carbon–oxygen bonds towards exchange has to be considered in designing biochemical labelling studies. The carbon–oxygen single bonds of alcohols, esters and ethers are sufficiently stable for oxygen-18 to be used as a label, but the carbonyl group is more problematic. Under acid catalysis, the carbonyl oxygen of an aldehyde such as benzaldehyde exchanges fairly rapidly. In this case, the exchange is facilitated by the presence of an adjacent aromatic ring stabilizing a carbocationic intermediate. On the other hand, ketones, even acetone, require longer and require either acid or base catalysis. The oxygen atoms of carboxylic acids and their salts are often sufficiently stable to enable them to be used as labelled substrates in aqueous media for biosynthetic experiments lasting several days. The carbonyl groups of amides and peptides are also relatively slow to exchange and hence these can be used in biochemical studies.

The introduction of an oxygen-18 label into an alcohol by the nucleophilic displacement of a halide or toluene-*p*-sulfonate is a simple method of synthesis. For example, displacement of the 5'-toluene-*p*-sulfonate of 2',3'-isopropylideneuridine with sodium [$^{18}$O]acetate gave, after hydrolysis of the acetate, 2,3-isopropylidene[5'-$^{18}$O]uridine (**7.28**).

**7.28**

Many of the methods that have been used to introduce an oxygen-18 label into an organic compound involve a hydrolytic process. In devising a synthetic scheme, it is important to consider carefully the origin of the oxygen atoms. The stepwise hydrolysis of a nitrile with [$^{18}$O]water *via* an imino ether provides a method for labelling the carbonyl oxygen of an ester. The first step involves the addition of an alcohol (ROH) to give the imino ether (RO–C=NH). The second step using the [$^{18}$O]water leads to the replacement of the imine by a carbonyl group in which the oxygen atom has come from the water. The subsequent reduction of the ester to an alcohol with lithium aluminium hydride leads to the formation of the

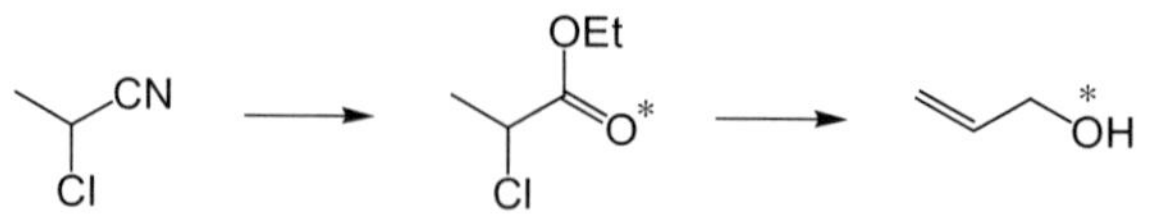

**Scheme 7.8**   The synthesis of allyl [$^{18}$O]alcohol.

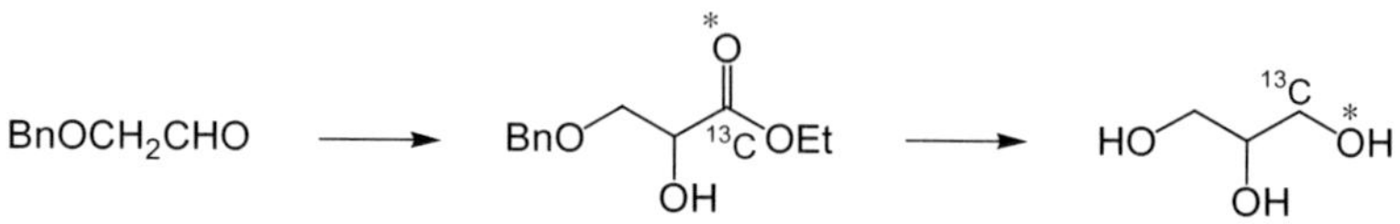

**Scheme 7.9**   The synthesis of [1-$^{13}$C,$^{18}$O]glycerol.

$^{18}$O-labelled alcohol corresponding to the acid of the ester. It is important to note the origin and the fate of the different oxygen atoms of an ester in this sequence. Thus allyl [$^{18}$O]alcohol (**7.31**) was prepared from 2-chloropropionitrile (**7.29**) by conversion of the nitrile to the carbonyl-labelled ester **7.30** *via* the imino ether (Scheme 7.8). Reduction of the ester afforded the labelled alcohol and elimination then gave allyl [$^{18}$O]alcohol (**7.31**).

Glycerol is a widespread primary metabolite. [1-$^{13}$C,$^{18}$O]Glycerol (**7.34**), in which the carbon-13 is acting as an NMR reporter label for the presence of the oxygen-18, was prepared by a sequence utilizing the hydrolysis of a nitrile (Scheme 7.9). Benzyloxyacetaldehyde (**7.32**) was prepared by ozonolysis of 1,4-dibenzyloxybut-2-ene. The carbon-13 label was then introduced from potassium [$^{13}$C]cyanide as the cyanohydrin. The cyano group was converted to its imino ether with acetyl chloride and ethanol. On hydrolysis with [$^{18}$O]water, only the carbonyl oxygen is labelled in **7.33** and this is the oxygen which is retained when the ester is reduced with lithium aluminium hydride. The benzyl protecting group was then removed by hydrogenolysis to give [1-$^{13}$C,$^{18}$O]glycerol (**7.34**).

Just as [$^{15}$N]benzylamine is a useful nucleophile for introducing nitrogen-15, so the alkoxide derived from benzyl [$^{18}$O]alcohol represents a convenient protected oxygen nucleophile. Once nucleophilic substitution has taken place, the benzyl protecting group can be removed by hydrogenolysis leaving the oxygen-18 label behind. The use of the benzyl protecting group ensures that only a single nucleophilic substitution takes place and prevents the formation of symmetrical ether by-products. Benzyl [$^{18}$O]alcohol is prepared by hydrolysis of the imino ether

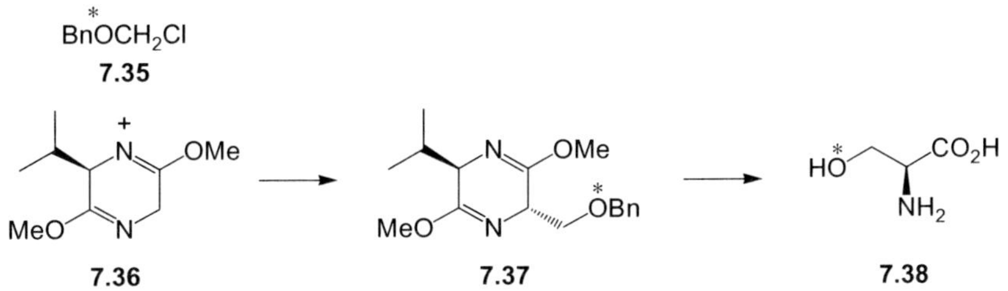

**Scheme 7.10**   The synthesis of L-[$^{18}$O]serine.

obtained from benzonitrile with [$^{18}$O]water followed by reduction of the ester with lithium aluminium hydride.

The preparation of L-[$^{18}$O]serine (**7.38**) is an illustration of this methodology (Scheme 7.10). The chloromethyl ether of benzyl [$^{18}$O]alcohol (**7.35**) was prepared using formaldehyde and hydrogen chloride under anhydrous conditions to prevent exchange of the benzylic oxygen label. This was then used to alkylate the bis-lactim ether **7.36** of cyclo(D-valylglycine). The D-configuration of the valine moiety directed the alkylation of the glycine moiety to give **7.37** and create the L-stereochemistry of the serine. Hydrolysis of the cyclic peptide, hydrogenolysis of the benzyl group and hydrolysis of the ester generated the labelled serine **7.38**.

Although the nucleophilic displacement of a leaving group such as the toluene-*p*-sulfonate by the labelled hydroxide can be used to form oxygen-18-labelled alcohols, the substitution may be accompanied by elimination and ether formation. A different method involves the reaction of a Grignard reagent with oxygen-18. This has been used in the preparation of [$^{18}$O]methanol and [$^{18}$O]phenol. The reaction may involve first the formation of a peroxide (**7.39**), which then reacts with a second molecule of the Grignard reagent. [$^{18}$O]Phenols have also been prepared by the fusion of arylsulfonates with potassium [$^{18}$O]hydroxide and by the hydrolysis of diazonium salts.

Oxyanions such as the permanganate, chromate and sulfate anions undergo exchange on heating with [$^{18}$O]water. In a study of the mechanism of hydroxylation of alkenes such as oleic acid, the oxygen-18 label from a permanganate was shown to be incorporated into the diol. Apart from the mechanistic implications of this work, this provides a way of labelling diols and some of their oxidation products.

Hydrogen [$^{18}$O]peroxide has been prepared from [$^{18}$O]water by passing the labelled water vapour through an electric discharge and trapping the resulting hydrogen peroxide. The labelled hydrogen peroxide was used in model studies of the mechanism of catalase and to prepare [$^{18}$O]-*m*-chloroperbenzoic acid from *m*-chlorobenzoyl chloride. The labelled peracid was used to prepare [$^{18}$O]nitrosobenzene by the oxidation of aniline but could also be used for epoxidation reactions.

The study of the mechanism of phosphoryl transfer reactions has made considerable use of oxygen-labelled phosphate esters. [$^{18}$O]Oxygen which is directly bonded to phosphorus gives rise to isotope shifts in the $^{31}$P NMR spectrum and the magnitude of these shifts reflect the phosphorus–oxygen bond order, A simple procedure for preparing labelled phosphate has been based on the hydrolysis of phosphorus trichloride with [$^{18}$O]water to give [$^{18}$O$_3$]phosphorus(III) acid. Methylation with diazomethane and treatment with chlorine gave the chlorophosphate **7.40**. Displacement, of the chlorine with a 2,4-dinitrophenyl group and subsequent removal of the methoxyl groups by C–O bond cleavage with bromotrimethylsilane gave a 2,4-dinitrophenyl ester (**7.41**) for mechanistic studies which was labelled in the non-bridging phosphate atoms.

Stereochemical studies with chiral [$^{16}$O,$^{17}$O,$^{18}$O]phosphate esters have shed valuable light on the mechanism of enzyme-catalysed phosphoryl transfer reactions. The synthetic route (Scheme 7.11) involved the separate introduction of the labels from [$^{17}$O]- and [$^{18}$O]water. The source of the chirality was the hydroxy acid (*S*)-mandelic acid (**7.42**). The mandelic acid was converted to the chiral benzoin by reaction with phenyllithium. The carbonyl group of the benzoin was labelled by hydrolysis of its ethylene ketal derivative with [$^{18}$O]water. Reduction with sodium borohydride gave [$^{18}$O]mesodihydrobenzoin. [$^{17}$O]Phosphorus oxychloride was prepared by the partial hydrolysis of phosphorus pentachloride with [$^{17}$O]water. Reaction of the labelled phosphorus oxychloride with the [$^{18}$O]mesodihydrobenzoin in the presence of methanol in pyridine gave

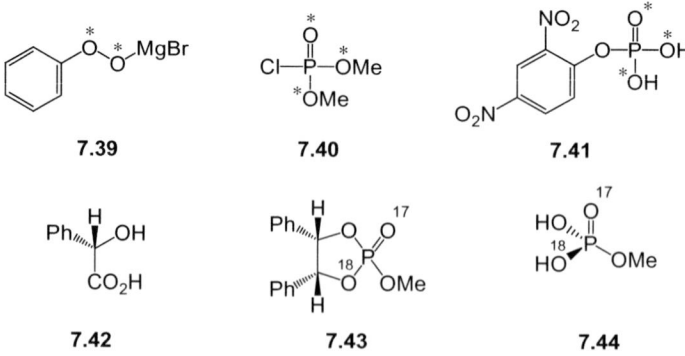

**Scheme 7.11**    The synthesis of chiral [$^{16}$O,$^{17}$O,$^{18}$O]phosphate.

[2-$^{17}$O]oxo[1-$^{18}$O]-1,3,2-dioxaphospholane (**7.43**), from which the chiral phosphate monoester **7.44** was obtained by hydrogenolysis.

The synthesis of carboxyl-labelled [$^{18}$O]amino acids may be achieved by the acid-catalysed hydrolysis of the esters with [$^{18}$O]hydrochloric acid. The acid-catalysed exchange of a carbonyl group has also been used in the preparation of [2-$^{18}$O]glycerol. Dihydroxyacetone was treated with [$^{18}$O] water and hydrogen chloride under otherwise dry conditions. The resulting [2-$^{18}$O]dihydroxyacetone was then reduced with sodium borohydride in dry methanol to give [2-$^{18}$O]glycerol.

As exemplified in the preparation of the labelled benzoin described above, the acid-catalysed hydrolysis of acetals and ketals provides a useful method for introducing an oxygen-18 label into a carbonyl group.

The oxygen atoms at C-3 and C-5 of the terpenoid precursor mevalonolactone have been labelled with oxygen-18. The [3-$^{18}$O]mevalonolactone **7.45** was prepared by a Reformatsky reaction between ethyl bromoacetate and a sample of 4-acetoxybutan-2-one which had undergone acid-catalysed exchange with [$^{18}$O]H$_2$O. The [5-$^{18}$O]mevalonolactone was prepared by the acid-catalysed hydrolysis of the acetal methyl 5,5-dimethoxy-3-hydroxy-3-methylpentanoate (**7.45**) in [$^{18}$O]H$_2$O. The acid catalyst was neutralized and the [$^{18}$O]aldehyde was reduced with sodium borohydride to give methyl [5-$^{18}$O]mevalonate (**7.45**). The lactone, containing 95% oxygen-18 at C-5 was obtained by acid-catalysed lactonization. There was some additional label at C-1.

**7.45**          **7.46**

Biosynthetic studies with sodium [1-$^{13}$C,$^{18}$O]acetate have been used to establish the origin of the oxygen atoms in polyketide biosynthesis. The $^{18}$O isotope shift in the $^{13}$C NMR spectrum provided the means of locating the site of the label. The doubly labelled acetate was obtained by heating sodium [1-$^{13}$C]acetate with [$^{18}$O]water containing hydrogen chloride followed by basification with sodium hydride.

CHAPTER 8

# Labelling with Isotopes of Phosphorus, Sulfur and the Halogens

There are both stable and radioactive isotopes of phosphorus, sulfur and the halogens which have been used for labelling purposes. In this chapter, we consider methods that have been used to prepare labelled compounds containing these isotopes.

## 8.1 PHOSPHORUS-32

The radioactive isotope of phosphorus, $^{32}$P, with a half-life of 14.2 days, has mainly found application in studies on phosphate metabolism. Phosphorus-32 can be prepared by the neutron irradiation of phosphorus-31 by the nuclear reaction $^{31}$P(n, $\gamma$)$^{32}$P. If this is carried out with red phosphorus, the product can be chlorinated easily to give phosphorus trichloride, from which various organophosphorus compounds can be obtained. In this context, it is worth noting that some radioactive compounds of phosphorus can be prepared by direct neutron irradiation. This has been used in the preparation of [$^{32}$P]nucleotides such as adenosine triphosphate. Another source of phosphorus-32 is the nuclear reaction of sulfur-32 $^{32}$S(n, p)$^{32}$P on neutron irradiation. In this case, the phosphorus is separated as a phosphate. The [$^{32}$P]phosphate of sodium [$^{32}$P]phosphate or phosphoric acid can be precipitated as calcium [$^{32}$P]phosphate and converted to [$^{32}$P]phosphorus oxychloride by treatment with phosgene.

Although chemical methods have been used, the preparation of many of the [$^{32}$P]nucleotides has been carried out enzymatically. Among the chemical methods that have been used are esterifications with 2-cyanoethyl

The Organic Chemistry of Isotopic Labelling
By James R. Hanson
© James R. Hanson 2011
Published by the Royal Society of Chemistry, www.rsc.org

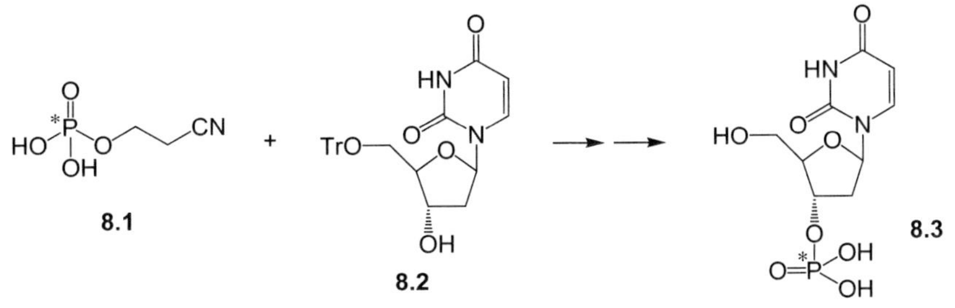

**Scheme 8.1** The synthesis of the 3'-[$^{32}$P]monophosphate of thymidine.

phosphate and dibenzyl phosphorochloridate (Scheme 8.1). 2-Cyanoethyl [$^{32}$P]phosphate (**8.1**) was prepared from [$^{32}$P]phosphorus oxychloride and 2-cyanoethanol. It was then coupled to an alcohol, for example in a sugar, with dicyclohexylcarbodiimide. This methodology has been used to prepare labelled nucleotides. For example, the free 3-hydroxyl of the protected deoxyribose of 5'-*O*-tritylthymidine (**8.2**) was coupled with 2-cyanoethyl [$^{32}$P]phosphate (**8.1**). The cyanoethyl protecting group is used because it is easily eliminated from the phosphate as acrylonitrile at the end of the sequence. After the coupling, hydrolysis of the trityl protecting group and elimination of acrylonitrile from the cyanoethyl ester afforded the 3'-mono[$^{32}$P]phosphate **8.3**. Dicyclohexylcarbodiimide in a dimethyl sulfoxide–pyridine mixture has been used to couple [$^{32}$P]phosphoric acid with nucleoside mono- and diphosphates and also with thiamin diphosphate to give the respective [γ-$^{32}$P]triphosphates.

Enzymatic methods have been used to transfer the labelled phosphate units to other alcohols. Adenosine 5'-[$^{32}$P]monophosphate has been obtained by incubating adenosine and adenosine triphosphate with a yeast extract containing [$^{32}$P]phosphate as the source of the label. The adenosine [$^{32}$P]monophosphate had to be separated chromatographically from inactive adenosine diphosphate. The terminal phosphates of ADP and ATP undergo exchange with inorganic [$^{32}$P]phosphate.

[$^{32}$P]Phosphorus oxychloride has been used for the phosphorylation of *N$^6$,O$^{5'}$*-ditrityladenosine. After removal of the trityl protecting groups with acetic acid, the product was separated chromatographically to give adenosine [2'-$^{32}$P]- and [3'-$^{32}$P]phosphates.

Phosphorus trichloride can be used for the introduction of a phosphorus-32 label into organophosphonates. An example involves the synthesis of ethyl[$^{32}$P]phosphonic acid. The [$^{32}$P]phosphorus trichloride was prepared by reducing [$^{32}$P]phosphorus oxychloride with triphenylphosphine. The ethyl[$^{32}$P]phosphonic acid was then obtained by heating ethyl bromide with [$^{32}$P]phosphorus trichloride and aluminium trichloride followed by hydrolysis with hydrochloric acid.

## 8.2  SULFUR-34 AND SULFUR-35

Sulfur has two isotopes that have been used in labelling studies. The stable isotope sulfur-34 has a natural abundance of 4.2% whereas the radioactive isotope sulfur-35 has a half-life of 87.2 days and emits β-radiation with a similar energy to that of carbon-14. A minor isotope, sulfur-33, which has a natural abundance of 0.76%, has a nuclear spin of 3/2 and can be observed by nuclear magnetic resonance spectroscopy.

Sulfur-35 is usually obtained by the neutron irradiation of potassium chloride in a nuclear reactor by the reaction $^{35}Cl(n, p)^{35}S$. The sulfur is separated as the sulfate. The sulfur in barium [$^{35}S$]sulfate can be converted to barium [$^{35}S$]sulfide by reduction with carbon or hydrogen at 900 °C, from which hydrogen [$^{35}S$]sulfide may be obtained by treatment with acid. The [$^{35}S$]sulfide may be oxidized to [$^{35}S$]sulfur with iodine and potassium iodide. [$^{35}S$]Sulfur dioxide has been obtained by the reaction of barium [$^{35}S$]sulfate with red phosphorus and oxygen.

The use of sulfur-35 as a tracer in biochemical studies has been relatively restricted. The labelling of thiols and thio ethers can be exemplified by the syntheses of the amino acids cysteine and methionine. The sulfur-35 is introduced by a nucleophilic substitution using benzene[$^{35}S$]thiol in which the benzyl group serves as a useful protecting group. Benzene[$^{35}S$]thiol has been prepared by the nucleophilic substitution of benzyl chloride with potassium hydrogen[$^{35}S$]sulfide or better by the reaction of benzylmagnesium chloride with sulfur-35. A suspension of sulfur in xylene for this purpose was obtained by oxidizing the radioactive sulfide ion with iodine in potassium iodide.

Benzyl[$^{35}S$]cysteine (**8.6**) and the dimer [$^{35}S$]cystine were prepared from β-chloroalanine (**8.4**) by coupling with [$^{35}S$]benzylthiomagnesium chloride (**8.5**) (Scheme 8.2). [$^{35}S$]Methionine (**8.8**) was obtained by reacting sodium [$^{35}S$]benzylthiolate with 1,2-dichloroethane to give the chloroethyl derivative **8.7**. This was used to alkylate diethyl phthalimidomalonate and

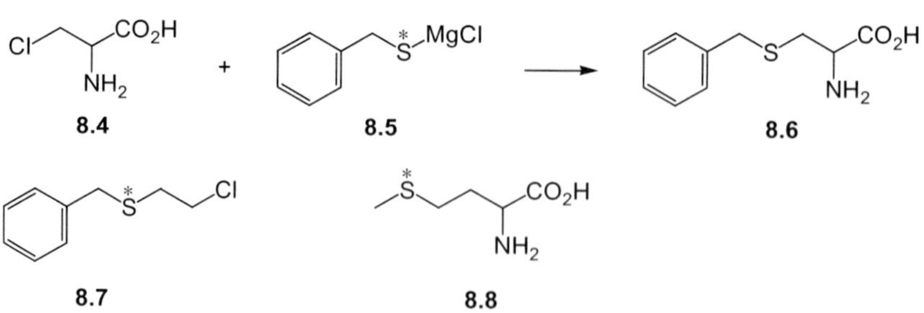

**Scheme 8.2**   The synthesis of benzyl[$^{35}S$]cysteine and [$^{35}S$]methionine.

the product was then hydrolysed to give a benzylthioamino acid. Hydrogenolysis of the benzyl group and methylation led to [$^{35}$S]methionine (**8.8**). The $^{35}$S-labelled amino acids have been incorporated into peptides such as [$^{35}$S]glutathione.

A more recent method for incorporating sulfur-35 into the amino acids methionine and cysteine and into the tripeptide glutathione is to grow a strain of baker's yeast, *Saccharomyces cerevisiae*, using sodium [$^{35}$S]sulfate as the sulfur source. Chromatographic fractionation of the protein hydrolysate gave the labelled amino acids.

A number of sulfoxides, sulfones and sulfonamides have been prepared using [$^{35}$S]thionyl chloride. Arylsulfonyl chlorides have also been prepared by the reaction of aryl Grignard reagents with elemental [$^{35}$S]sulfur followed by oxidation and chlorination.

The sulfur of a thiourea can be exchanged by reaction with sulfur-35 in refluxing pyridine. This was used in the preparation of [6-$^{35}$S]thioguanosine for use in the study of ribonucleic acids.

Sulfur-34 is a stable isotope of sulfur whose presence can be detected by mass spectrometry. [$^{34}$S]Sulfur dioxide has been obtained at 25% concentration by exchange of sulfur dioxide with sodium hydrogensulfite. Sulfur-34 is also available in the elemental form. Although it has not been widely used, the observation of an isotope effect in the equilibration of the bisulfite adducts of aldehydes and ketones with sodium hydrogensulfite was used to confirm the presence of a C–S bond in their structures. The significance of the isotope effect was assessed by comparing it with the minimal effect that was observed on the hydrolysis of sulfate esters. The mechanistic implications of sulfur-34 isotope effects in sulfate ester hydrolysis have also been examined. A large $^{34}$S kinetic isotope effect was observed in acid-catalysed hydrolysis, suggesting that cleavage of the S–O bond takes place during the rate-limiting step.

Although there are many rearrangements in the organic chemistry of sulfur compounds, relatively few have been explored by isotopic labelling. One recent example is the establishment of the intramolecular nature of a lithium diethylamide-mediated rearrangement of an aryl triflate (ArOSO$_2$CF$_3$) to an aryltriflone [Ar(OH)SO$_2$CF$_3$] using evidence based on the lack of any cross-over between an isotopomer labelled with deuterium in the aromatic ring and another labelled with sulfur-34 in the triflate. The sulfur-labelled triflate was prepared from elemental sulfur-34. The sulfur was reacted with potassium cyanide to give potassium [$^{34}$S]thiocyanate, which was used to prepare benzyl [$^{34}$S]thiocyanate from which benzyl trifluoromethyl[$^{34}$S]thio ether was obtained. Oxidation of the latter with chlorine and water gave trifluoromethyl[$^{34}$S]sulfonyl chloride, which was employed to give the required aryl [$^{34}$S]triflate.

Small but significant isotope effects have been observed in the metabolism of sulfur in biological systems leading to variations in sulfur isotope ratios. The isotope fractionation of sulfur by bacteria which reduce sulfate to sulfide and are involved in sulfite disproportionation has been used to interpret the fate of sulfur in the geological record. Sulfur isotope ratios have been used in the discussion of the dynamics of sulfur metabolism in ecosystems and in characterizing pollutants in water. Isotope ratios have also been examined in geothermal deposits of sulfur.

## 8.3  CHLORINE-36

Chlorine-36 has a long half-life ($3.01 \times 10^5$ years) and has been relatively little used. It is formed along with sulfur-35 in the neutron irradiation of potassium chloride by the nuclear reaction $^{35}Cl(n, \gamma)^{36}Cl$. The sulfur-35 which is formed at the same time is removed and the potassium chloride is recycled in order to obtain a reasonable conversion. The potassium [$^{36}$Cl]chloride may be converted to [$^{36}$Cl]chlorine, from which a number of chlorinating reagents (*e.g.* thionyl chloride) have been obtained. Potassium [$^{36}$Cl]chloride has been used in biosynthetic studies with fungal metabolites to establish the origin of chlorine substituents. Sodium chloride enriched to 90% $^{35}Cl$ is available.

## 8.4  IODINE ISOTOPES

Iodine has a number of isotopes which have been used for different purposes. Iodine-123, which has a short half-life, is used for single photon emission computed tomography (SPECT) and is discussed in Chapter 9. Iodine-125 is produced by the neutron bombardment of xenon-124 in the reaction $^{124}Xe(n, \gamma)^{125}Xe \rightarrow ^{125}I$. The xenon-125, which is formed first, decays with a half-life of 18 h to iodine-125, which has a half-life of 59.9 days. Iodine-125 has also been produced in a cyclotron by bombardment of tellurium-125 with deuterons, $^{125}Te(d, 2n)^{125}I$. Iodine-125 decays by the emission of $\gamma$-rays and X-rays and hence some shielding is required in order to work with this isotope. The effect of this radiation on tissue accounts for one at the uses of the isotope, and that of iodine-131, in radiation therapy of some tumours such as prostate cancer. Iodine-131 has been used in the treatment of thyroid disease, because of the tendency of iodine to concentrate in the thyroid gland, particular precautions should always be taken when working with iodine isotopes.

Radioactive iodine is introduced into organic compounds by halogen exchange, by direct iodination with iodine or iodine monochloride, by the Sandmeyer displacement of the diazo group, by iododestannylation and occasionally by a biosynthetic method.

A major use of iodine-125 is in radioimmunoassay. This analytical technique is based on the specific and reversible formation of an antigen (hapten)–antibody complex. In the first stage of this analysis, a labelled antigen such as a peptide hormone is bound to a specific antibody. In the calibration stage, known amounts of the unlabelled antigen are equilibrated with the labelled antigen–antibody complex. Some of the labelled antigen is displaced. The bound and free radioactivity can then be measured. A binding curve can be created in which the ratio of the bound to free labelled antigen is plotted against the amount of added unlabelled antigen. The unknown amount of antigen in, for example a blood sample can then be established by measuring the bound to free isotope ratio and comparing it with the binding curve. Iodine-125 is a useful isotope for this work because sodium [$^{125}$I]iodide in the presence of an oxidizing agent will readily react with the amino acids tyrosine and histidine, which occur in many peptides, and iodinate them within the intact peptide. The radioiodination procedure involves oxidizing the alkaline sodium iodide to the iodonium ion in the presence of chloramine-T or a peroxidase and hydrogen peroxide. The method can be used to detect nanomolar amounts of a hormone in a blood sample. Among the many methods that have been developed are protocols for the analysis of insulin in diabetes, angiotensin II in heart failure, gastrin in patients with peptic ulcer and the detection of growth hormone. More recently, the non-radioactive ELISA (enzyme-linked immunosorbent assay) methods have been used for these analytical purposes.

Some iodine-125-labelled compounds have been prepared for specific purposes. These include [$^{125}$I]-5-iodo-2-deoxyuridine (**8.9**) as an analogue for the nucleic acid base thymidine. The preparation was based on the reactivity of C-5 of the pyrimidine ring system towards electrophiles.

**8.9**

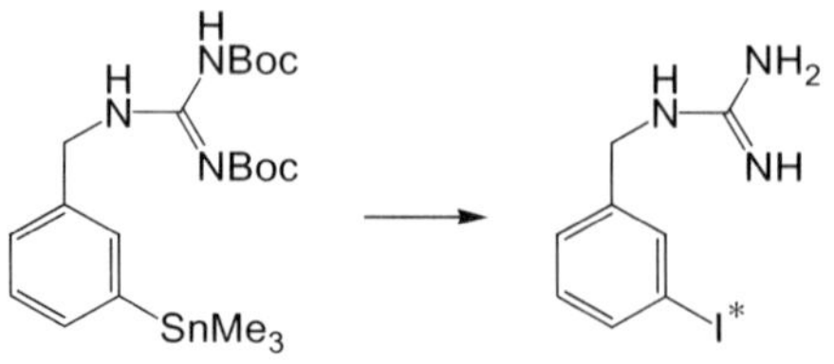

**8.10**                                    **8.11**

**Scheme 8.3**   Iododestannylation in the synthesis of *m*-iodobenzylguanidine.

Iodine-131, which has a much shorter half-life (8 days), is formed along with iodine-129 in the nuclear decomposition of uranium. It is still used to detect and treat thyroid disease. To be of value in therapy, the emission from the radioisotope must excite electrons particularly in water to produce hydroxyl free radicals. These may then react with adjacent biologically active molecules and damage the growing tissue. [$^{131}$I]-*m*-iodobenzylguanidine (MIBG) (**8.11**) is used in the treatment of some tumours of the adrenals and the nervous system. The iodine is introduced on to the aromatic ring by iododestannylation on  *N,N'*-bis(*tert*-butyloxycarbonyl)-3-trimethylstannylbenzylguanidine (**8.10**) (Scheme **8.3**).

[$^{131}$I]Iodoacetamide and *N*-[$^{131}$I]iodoacetylamino acids have been prepared by the displacement of chlorine or bromine from chloro- or bromacetyl derivatives with sodium [$^{131}$I]iodide. The labelling of thyroxine (**8.12**) has been achieved by an exchange reaction involving heating a solution of thyroxine in an alcohol with sodium [$^{131}$I]iodide. Diiodotyrosine was prepared by iodination of tyrosine with a solution of [$^{131}$I]iodine prepared by the oxidation of sodium [$^{131}$I]iodide with a mixture of potassium iodide and potassium iodate. The carcinogenic 2-acetylamino-7-[$^{131}$I]iodofluorene (**8.13**) was obtained by the iodination of 2-acetyl-aminofluorene using [$^{131}$I]iodine chloride. The iodine of the iodine chloride underwent exchange with sodium [$^{131}$I]iodide to form the labelled reagent. Decomposition of diazonium salts using sodium [$^{131}$I]iodide affords labelled aryl iodides. Thus [$^{131}$I]iodobenzene was obtained from aniline and used to prepare the [$^{131}$I]iodo analogue of DDT by condensation with trichloroacetaldehyde in the presence of chlorosulfonic acid.

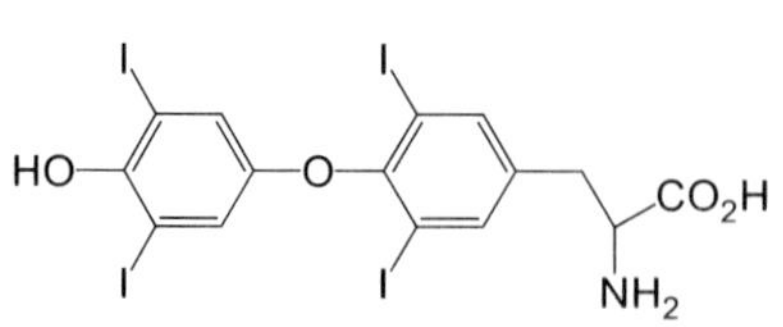

**8.12** 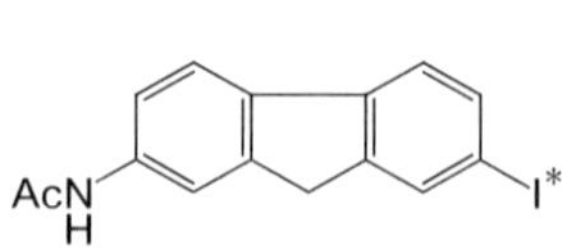                                    **8.13**

CHAPTER 9

# Labelling Organic Compounds for Diagnostic Imaging

## 9.1 INTRODUCTION

Drugs containing a radioactive nuclide are often described as radio-pharmaceuticals. Their medical applications lie in both diagnosis and therapy. Diagnostic radiopharmaceuticals are molecules that are typically labelled with a positron-emitting isotope for use in positron emission tomography (PET) or a gamma-emitting isotope for use in single photon emission computed tomography (SPECT). Diagnostic radiopharmaceuticals are designed to target particular receptors or organs and to provide a non-invasive method of assessing the function of the target and by using isotopes with a short half-life, with relatively little radiation damage to the subject. Therapeutic radiopharmaceuticals, on the other hand, are designed to deliver a cytotoxic radiation dose to diseased, typically tumour, cells. The best known of these are compounds containing radio-iodine, which are used in the treatment of thyroid disease and which were discussed in the previous chapter. Target specificity is again required to minimize radiation damage to healthy tissue.

## 9.2 POSITRON EMISSION TOMOGRAPHY

The development of PET has had a major impact on the synthesis of labelled compounds, requiring the development of rapid methods for

The Organic Chemistry of Isotopic Labelling
By James R. Hanson
© James R. Hanson 2011
Published by the Royal Society of Chemistry, www.rsc.org

incorporating short-lived isotopes such as carbon-11 and fluorine-18 into compounds. Although carbon-11 was described in the 1930s, its short half-life of 20.4 min imposed constraints on its use as a radio-tracer and it was soon overshadowed by the use of carbon-14. Both carbon-11 and fluorine-18 (half-life 109.7 min) decay by the emission of a positron (a $\beta^+$ particle). The positrons from fluorine-18 have a relatively low energy (0.635 MeV) and are annihilated by collision with an electron close (0.5–2 cm) to their site of emission. The annihilation releases two photons at 180° to each other, which can be detected and their point of origin calculated. The higher the energy of the positron, the further it travels before annihilation occurs and the more diffuse any calculated image appears. To be of value in tomography, the photons that are produced in the annihilation event must be of sufficiently high energy to escape the body in order to be observed. The detection of these and the identification of their point of origin enable the location of the parent isotope to be defined. Hence by computation it is possible to create an image of the tissue in which compounds containing these isotopes are located and are being metabolized. In this way, it is possible to image tumours and other groups of cells such as those in parts of the brain and in the heart.

The short half-life of these isotopes becomes an advantage because it minimizes the radiation damage to the living tissue. Nevertheless, the short half-life also dominates the chemistry of the isotope and the way in which it is incorporated into a biological molecule. For most purposes, less than three half-lives must elapse between the end of the bombardment leading to the formation of the isotope and it being used in the clinic. For carbon-11, this means that the synthesis and purification must be completed in less than 1 h to allow for a convenient start of its *in vivo* use. The incorporation of fluorine-18 has a slightly longer synthetic time-scale. The need for rapid radio-tracer synthesis has led to the modification of many of the reactions of stable isotope chemistry. Microwave methods have played a significant role in reducing reaction times. Furthermore, the reactions must be carried out on the micromolar scale under conditions of radiation protection and with only short HPLC purification steps. Some specialized techniques and apparatus have been developed to handle the very small amounts of material involved in very rapid reactions. Since the product may be administered intravenously in the clinic, purity and sterility are very important. It is also worth noting that the use of a racemate of a chiral compound may produce problems from the background emission arising from the partial metabolism or transport of the biologically-inactive enantiomer.

## 9.3 CARBON-11 IN PET

Carbon-11 is produced by the bombardment of nitrogen gas with high-energy protons in a cyclotron. The nuclear reaction $^{14}N(p, \alpha)^{11}C$ takes place with the emission of an alpha particle. Small amounts of oxygen in the target allow the carbon-11 to react to form $[^{11}C]$carbon dioxide from which further single carbon species are generated. Alternatively, the presence of small amounts of hydrogen lead to the formation of $[^{11}C]$methane. The labelling chemistry then involves introducing the carbon-11 units rapidly into the target molecule as near as possible to the end of any synthetic sequence. In this context, it is not just the time spent on the reactions with the labelled substrate but also that spent on the purification that needs to be minimized.

The Grignard reaction provides a rapid means of introducing carbon-11 into a carboxylic acid. If the carbon-11 in $[^{11}C]$carbon dioxide is trapped by a Grignard reaction as $[1-^{11}C]$acetate, this can then be used to acetylate a substrate.

The selection of the site for introducing a label is of paramount importance in the rapid labelling of compounds for use in PET. This is exemplified by the further transformation of a carboxyl group in the preparation of $[^{11}C]$diprenorphine (**9.1**) as a ligand for imaging the opiate receptor in the brain. The Grignard reaction between cyclopropylmagnesium bromide and $[^{11}C]$carbon dioxide was completed in 2 min. The acyl chloride was obtained from the Grignard salt by chlorination with phthaloyl chloride and reacted with a diprenorphine precursor. Rapid reduction with lithium aluminium hydride released the phenol and reduced the amide to give **9.1**. The overall sequence was completed in 53 min from the end of bombardment. This sequence shows how a complicated polyfunctional molecule can be efficiently labelled in a short time.

Apart from routes using acyl chloride, $[^{11}C]$amides have been prepared in a rapid reaction ($\sim 30$ min) from the magnesium $[^{11}C]$carboxylates, an amine and excess alkylmagnesium halide. The $[^{11}C]$carbon dioxide which is produced in the cyclotron is trapped by the organomagnesium halide. The magnesium carboxylate is then treated with the amine in the presence of excess alkyl magnesium halide.

Brief treatment of the $[^{11}C]$carbon dioxide with zinc at 400 °C generates $[^{11}C]$carbon monoxide. Specific trapping systems have been designed to concentrate this gas. The carbonylation of organohalides with a zerovalent palladium reagent, typically $Pd(PPh_3)_4$, and a nucleophile has provided a versatile method of preparing carbon-11-labelled carboxylic acids, esters, amides, aldehydes and ketones. These have been particularly useful when combined with a transmetallation reaction using organotin chemistry. Thus $[1-^{11}C]$ethyl iodide has been prepared from methyl iodide and $[^{11}C]$carbon monoxide. These gave $[1-^{11}C]$acetic acid and methyl $[1-^{11}C]$acetate in a palladium-mediated carbonylation. They were immediately reduced with lithium aluminium hydride to $[1-^{11}C]$ethanol and converted to the iodide in an overall sequence lasting 20 min. [*carbonyl*-$^{11}C$]Acetophenone has been synthesized *via* a palladium-promoted coupling of iodobenzene, $[^{11}C]$carbon monoxide and tetramethyltin to provide the carbon nucleophile. $[2-^{11}C]$Acetone was prepared using methyl iodide as the halide in this reaction.

$[^{11}C]$Amides have been prepared with primary and secondary amines as the nucleophiles and subjected to further transformations. For example, [*carbonyl*-$^{11}C$]-*N*-benzylbenzamides have been prepared from an aryl iodide, benzylamine and $[^{11}C]$carbon monoxide. In this case, the catalyst was palladium dichloride on silica. The value of these carbonylation reactions has also been enhanced by trapping the initial $[^{11}C]$carbon monoxide as its borane complex ($BH_3 \cdot CO$), reducing the need to work under pressure. A rhodium-catalysed carbonylation of ethyl diazoacetate in the presence of ethanol has been developed for the rapid synthesis of diethyl [*carbonyl*-$^{11}C$]malonate.

$[^{11}C]$Carbon monoxide has been converted to $[^{11}C]$phosgene and this has been used in a rapid synthesis of $[2-^{11}C]$thymine (**9.4**) (Scheme 9.1). The precursor β-(*N*-benzoylamino)methacrylamide (**9.3**) was prepared from ethyl α-formylpropionate (**9.2**), ammonia and benzoyl chloride. This was then cyclized with $[^{11}C]$phosgene to form the pyrimidine ring. The benzoyl group was removed by rapid hydrolysis to give **9.4**.

Reduction of $[^{11}C]$carbon dioxide to $[^{11}C]$methanol and then iodination with hydrogen iodide afforded $[^{11}C]$methyl iodide, An alternative procedure involved the iodination of $[^{11}C]$methane in the gas phase. These rapid sequences allow methyl iodide to be used in a series of alkylation reactions. One example (Scheme 9.2) involves the preparation of L-$[3-^{11}C]$alanine. Alkylation of the chiral unit **9.5**, in which the bulky *tert*-butyl group directs the stereochemistry of the reaction, gave the

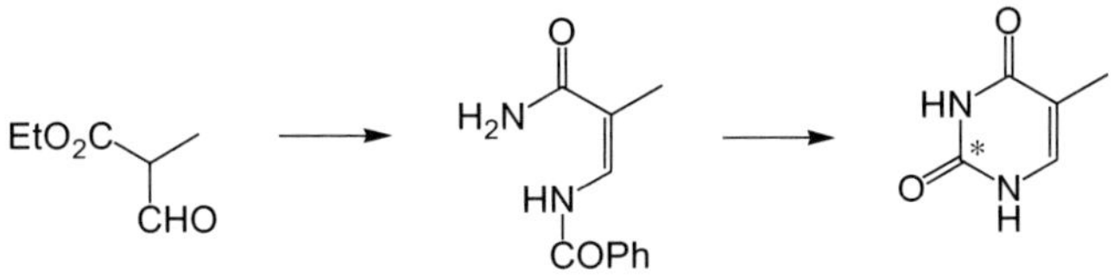

**Scheme 9.1**   The synthesis of [2-$^{11}$C]thymine.

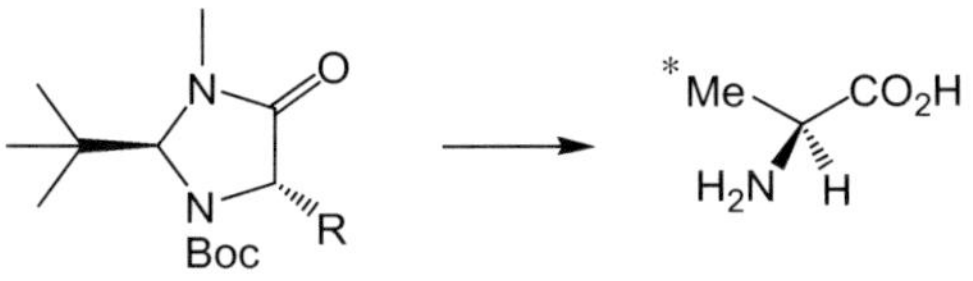

**9.5   R = H**                          **9.7**
**9.6   R = $\overset{*}{Me}$**

**Scheme 9.2**   The synthesis of L-[3-$^{11}$C]alanine.

amino acid derivative **9.6**. This was hydrolysed to afford L-[3-$^{11}$C]alanine
(**9.7**).

There have been a number of examples of the alkylation of heteroatoms. Thus alkylation on sulfur has been used to prepare L-[*methyl*-$^{11}$C]
methionine.

Imaging dopamine receptors in the brain with labelled cocaine analogues such as β-carboxymethyliodophenyltropane (β-CIT) has been an
important application of **PET**. Alkylation on nitrogen has been used to
prepare [*N-methyl*-$^{11}$C]-β-CIT (**9.8**). The dopamine transporter complex
in the brain is involved in the presynaptic reuptake of dopamine in
dopaminergic nerve cells and hence in the regulation of dopamine release. Cocaine analogues bind to these receptors and these compounds
have been labelled with carbon-11, fluorine-18 and iodine-123 for imaging these centres.

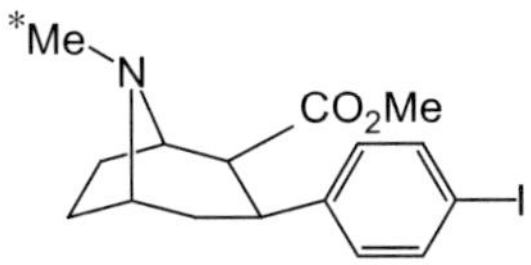

**9.8**

[$^{11}$C]Methyl triflate ([$^{11}$C]-MeOSO$_2$CF$_3$) is a useful alternative to
[$^{11}$C]methyl iodide for methylation reactions. It can be rapidly prepared

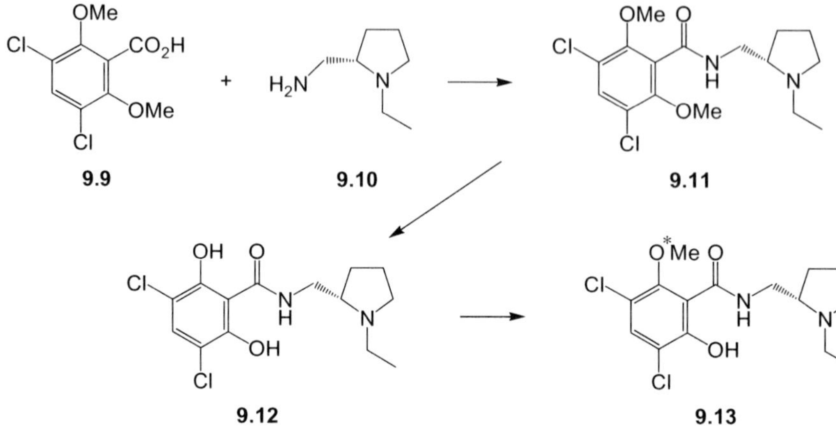

**Scheme 9.3** The synthesis of [$^{11}$C]raclopride.

from [$^{11}$C]methyl iodide by treatment with silver triflate. An example of its use is in the preparation of [$^{11}$C]raclopride (**9.13**) by methylation of a phenolic hydroxyl group in desmethylraclopride (**9.12**) (Scheme 9.3). Raclopride is an important ligand for imaging the dopamine D$_2$ receptor in the brain. The synthetic precursor was easily prepared in three steps from 3,5-dichloro-2,6-dimethoxybenzoic acid (**9.9**) and (*S*)-(−)-2-ami-nomethyl-1-ethylpyrrolidine (**9.10**) followed by demethylation of **9.11** to give **9.12**. The radiosynthesis step, purification and formation of the product for use as an imaging agent could be completed in 35 min.

The use of labelled steroids for imaging steroid receptors represents another potential application of PET. [$^{11}$C]Methyllithium has been used to prepare [26-$^{11}$C]vitamin D$_3$ (**9.14**) by addition to the corresponding nor-ketone. The labelled vitamin D$_3$ was required for mapping the vitamin D receptor.

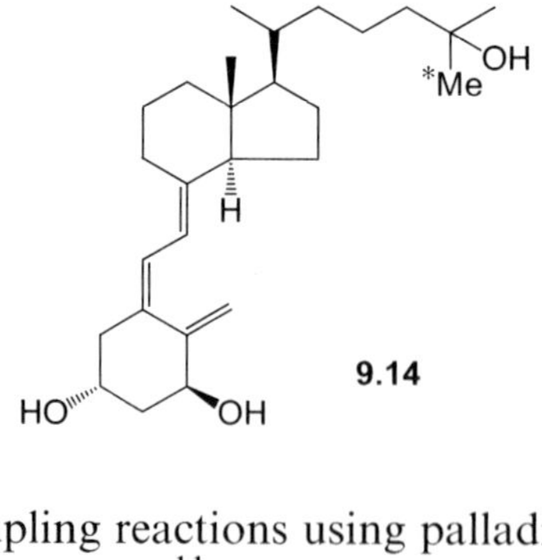

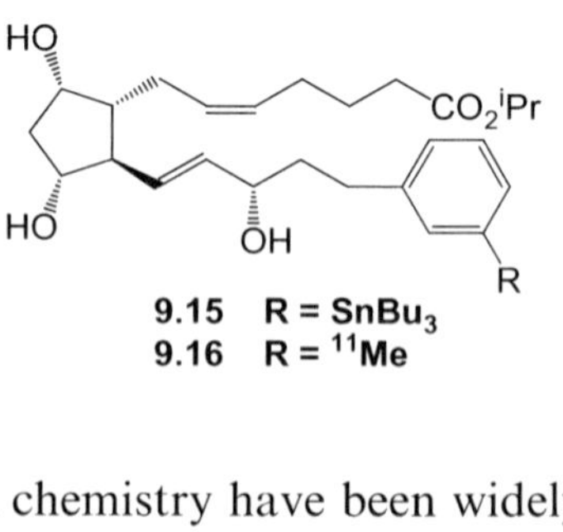

Coupling reactions using palladium chemistry have been widely used to introduce [$^{11}$C]methyl groups on to aromatic rings and other unsaturated systems. These include palladium(0) Stille coupling re-actions involving the displacement of organostannanes. For example, the preparation of the prostaglandin analogue 17-(3-[$^{11}$C]methylphe-nyl)-18,19,20-trinorPGF$_{2\alpha}$ isopropyl ester (**9.16**) used a Stille coupling of the organostannane **9.15** with [$^{11}$C]methyl iodide in the presence of tris(dibenzylideneacetone)dipalladium(0) and tri(*o*-tolyl)phosphine ac-tivated by copper(I) chloride and potassium carbonate. In another ex-ample, an [$^{11}$C]estradiol derivative (**9.17**) was prepared from the ethynylestradiol mestranol by a coupling reaction using [$^{11}$C]methyl iodide. Because of potential problems that might arise from traces of tin residues in compounds that might be used for medical imaging, the Suzuki coupling using arylboronic acids has been examined for some of these reactions.

**9.17**              **9.18**

Other reactive single-carbon species such as [$^{11}$C]nitromethane have been obtained from methyl iodide by nucleophilic displacement. The condensation of [$^{11}$C]nitromethane with pyrylium salts provided an entry into [$^{11}$C]benzenoid compounds, as exemplified by the synthesis of 4-nitroanisole (**9.18**) (*cf.* Chapter 2).

Heating [$^{11}$C]methane with nitrogen over a platinum catalyst provided a route to hydrogen [$^{11}$C]cyanide. A number of compounds were then prepared from the labelled cyanide. These include amino acids such as tyrosine by the Strecker synthesis and D-[1-$^{11}$C]glucose (**9.20**) together with D-[1-$^{11}$C]mannose from the C$_5$ sugar D-arabinose *via* reduction of the nitrile in [1-$^{11}$C]-D-aldonitrile (**9.19**) to an aldehyde (Scheme 9.4).

## 9.4 FLUORINE-18 IN PET

Fluorine-18 has attracted considerable interest in the context of PET because its short half-life of 110 min nevertheless allows time for syn-thetic labelling reactions. The positrons which are emitted have a sufficiently limited range in tissues before the photons are emitted as a result of annihilation, to give well-defined PET images. Fluorine-18 is prepared in the cyclotron by two processes. The first involves

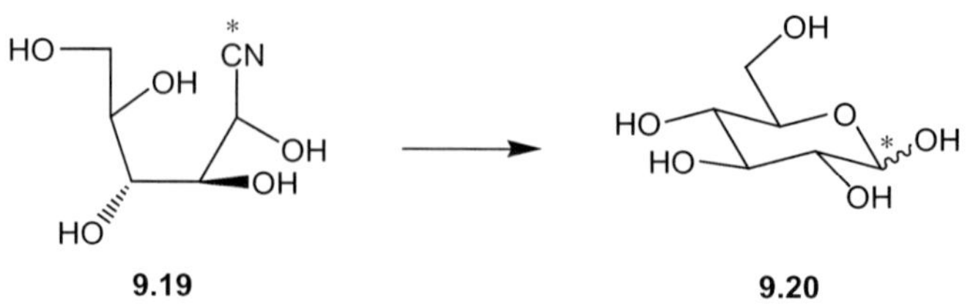

**Scheme 9.4**   The synthesis of [1-$^{11}$C]glucose.

bombardment of neon gas with deuterons to bring about the reaction $^{20}$Ne(d, $\alpha$)$^{18}$F, which leads to [$^{18}$F]fluorine gas. Because of the amounts involved, this is recovered by dilution using fluorine-19 as a carrier. In the light of the corrosive properties of fluorine gas, this method is not particularly attractive as a route for introducing fluorine-18 into a la-belling reaction. The second, more common, method involves the bombardment of [$^{18}$O]water to give the fluoride ion by the reaction $^{18}$O(p, n)$^{18}$F. This is trapped by ion-exchange methods and readily ob-tained as tetrabutylammonium fluoride.

The major reactions that have been used in the preparation of fluorine-18 compounds are based on the fluoride ion and involve the nucleophilic substitution of sulfonate esters or other halogens, the opening of an ep-oxide, the addition of a mixed halide (usually BrF) and the nucleophilic aromatic substitution of a leaving group such as a trimethylammonium substituent on a deactivated aromatic ring. Electrophilic aromatic sub-stitution reactions with fluorine-18 or acetyl hypofluorite are less widely used, not only because they may involve dilution with carrier fluorine-19 but also because they can result in a loss of half of the label. Fluoro-destannylation and fluorodemercuration reactions have been used. However, the replacement of a diazonium group by fluorine has not been widely used to introduce fluorine-18 on to an aromatic ring.

The need to complete the chemical synthesis in the shortest possible time dictates both the synthetic steps and the experimental methodology such as the choice of solvent. Aprotic solvents such as dimethyl sulfoxide and acetonitrile and cryptands (*e.g.* Kryptofix 222$^{\circledR}$) to complex the metal counter-ion are used to expose a 'naked' fluoride ion. A number of the reactions have been carried out with microwave irradiation.

One of the most widely used fluorine-18 labelled compounds is 2-[$^{18}$F]fluoro-2-deoxy-D-glucose ([$^{18}$F]FDG) (**9.22**). It is used to image glucose uptake by tissues. Rapidly growing tissues such as cancers take up the labelled glucose more rapidly than normal tissue and the site of fluorination, by blocking normal glycolysis, reduces the rate at which the

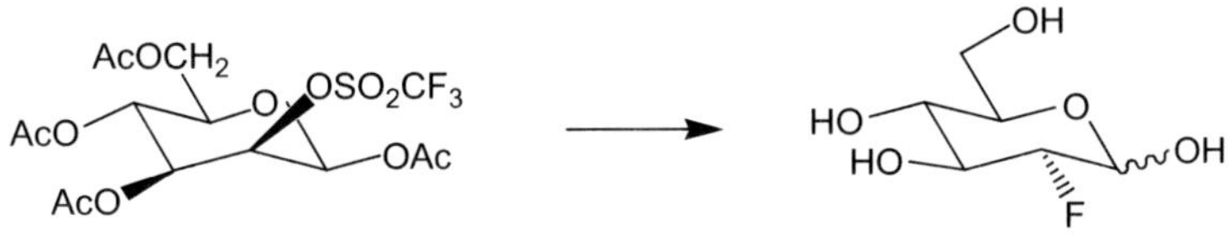

**Scheme 9.5** The synthesis of 2-[$^{18}$F]fluoro-2-deoxy-D-glucose.

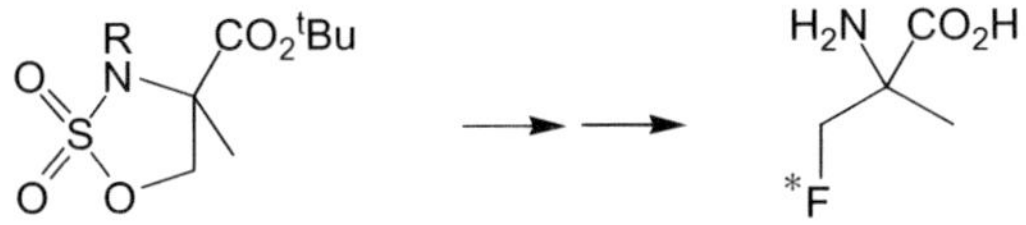

**Scheme 9.6** The synthesis of 2-amino-3-[$^{18}$F]fluoro-2-methylpropanoic acid.

labelled glucose is metabolized. As a result, these tissues can be imaged using PET. The original synthesis of [$^{18}$F]FDG (**9.22**) involved the electrophilic fluorination of D-glucal. However, the normal synthesis (Scheme 9.5) employs the nucleophilic substitution of the 2-O-trifluoro-methanesulfonate of tetraacetoxy-D-mannose (**9.21**) by an [$^{18}$F]fluoride ion. The reaction uses a 'naked' fluoride ion from potassium [$^{18}$F]fluoride in the presence of Kryptofix 222$^®$ in acetonitrile. An excess of the sugar is used and there can be problems of purity of the product. These have been obviated by attaching the precursor to the fluorodeoxyglucose to a solid support through a sulfonate linker. Displacement of the sulfonate linker by the fluorine-18 gave only [$^{18}$F]FDG (**9.22**).

Radiolabelled amino acids have also proved useful for imaging tumours, particularly in the brain. α-Methylamino acids such as 2-amino-3-fluoro-2-methylpropanoic acid (**9.24**) and its N-methyl analogue have been examined in this context. Whereas the unlabelled fluoroamino acid was prepared from fluoroacctone using a Strecker synthesis, the labelled amino acid was synthesized by a route in which the fluorine-18 was introduced at the end of the synthesis (Scheme 9.6) by displacement of a cyclic sulfamide. This involved the nucleophilic displacement of the sulfonyl group in the sulfamidate **9.23** with the [$^{18}$F]fluoride ion in the presence of a cryptand and potassium carbonate. Rapid hydrolysis of the protecting groups gave **9.24**. This particular synthesis illustrates the need for selectively adjusting the functionality and the protecting groups in a precursor in order to be able to introduce the fluorine label rapidly and efficiently in the last stages of a synthetic sequence.

Various fluorine-18-labelled steroids have been prepared by the nucleophilic substitution of sulfonate esters and used for imaging different steroid receptors. [7α-[18]F]Fluoro-17α-methyl-5α-dihydrotestosterone **(9.25)**, a ligand for imaging the androgen receptor in the study of prostate cancer, was prepared by the no-carrier-added displacement of the 7β-toluene-*p*-sulfonate with [[18]F]tetrabutylammonium fluoride. Similarly, the [16α- and 16β-[18]F]fluoroestradiols, which were required for imaging the estrogen receptor in the study of breast cancer, were prepared by the substitution of the respective trlfluoromethanesulfonates. The displacement of 16,17-cyclic sulfates was also used.

9.25                                    9.26

Fluorine-18-labelled cocaine analogues have been used for PET imaging dopaminergic centres in the brain. An example is 2β-carbo(1'-fluoro-2-propoxy)-3β-(4''-chlorophenyl)tropane **(9.26)**. Both enantiomers of the ester were prepared by a route in which the fluorine-18 was introduced in the last step by nucleophilic displacement of the 1'-methanesulfonate. Substituted [[18]F]fluoroalkanes such as 1-bromo-3-[[18]F]fluoropropane have been used for the introduction of *N*-fluoroalkyl groups. The fluorine-18 was introduced into disubstituted alkanes $X(CH_2)_nX$, where $n = 2$ or 3 and X = Br, OMs or OTs, from the [[18]F]fluoride ion and potassium carbonate in acetonitrile in the presence of the cryptand Kryptofix 222®. Thus, *N*-[[18]F](3'-fluoropropyl)-2β-carbomethoxy-3β-(4'-iodophenyl)tropane ([18]FP-β-CIT) **(9.27)** has been prepared from the corresponding nortropane by distilling the bromofluoropropane into a vessel containing the nortropane, diisopropylamine and potassium iodide. The *N*-alkylated product was obtained within 30 min.

9.27                                    9.28

The [$^{18}$F]halofluorination of alkenes is another method that has been used for the introduction of fluorine-18 into a compound. This may be exemplified by the preparation of [11β-$^{18}$F]-5α-dihydrotestosterone (**9.28**) for imaging androgen receptors in the study of prostate disease. A bromofluorination was carried out by treating the 9(11)-unsaturated steroid with the [$^{18}$F]fluoride ion, *N,N*-Dibromohydantoin and a small amount of concentrated sulfuric acid to give a 9α-bromo-11β-fluorosteroid. The 9α-bromine substituent was removed in a subsequent reduction.

Although it has been carried out, electrophilic aromatic fluorination is a difficult reaction in the context of introducing a fluorine-18 label on to an aromatic ring. Many [$^{18}$F]fluoroaromatic compounds have been prepared by the nucleophilic aromatic substitution of substituents on aromatic rings bearing electron-withdrawing substituents. [$^{18}$F]-α-Methyl-L-amino acids such as [$^{18}$F]-α-methyl-L-phenylalanine (**9.31**) have been used in imaging the brain in neurodegenerative diseases. The fluorine-18 substituent was introduced on to the aromatic ring by nucleophilic substitution of *o*-trimethylammoniumbenzaldehyde triflate (**9.29**) with a [$^{18}$F]fluoride ion in the presence of potassium carbonate and Kryptofix 222® (Scheme 9.7). The electron-withdrawing aldehyde activates the trimethylammonium group to nucleophilic substitution by the fluoride. The resulting *o*-[$^{18}$F]fluorobenzaldehyde (**9.30**) was then used in a rapid synthesis of the labelled α-methylamino acid **9.31**.

This methodology (Scheme 9.8) has been used to prepare [2-$^{18}$F]fluoroestradiol (**9.34**). The electron-rich aromatic ring of estradiol was deactivated by oxidation of the 3,17β-dimethyl ether to a C-6 ketone (**9.32**). The trimethylammonium group was then introduced at C-2 by nitration, selective reduction and methylation. In this case, the C-6 ketone activated the C-2 trimethylammonium group in **9.33** to nucleophilic substitution. Once the fluorine-18 label had been introduced by displacement of the trimethylammonium group, the C-6 carbonyl group and protecting methyl ethers were removed to give **9.34**.

The alkylated amino acid L-α-methyltyrosine has an affinity for certain brain tumours. The direct fluorination of L-α-methyltyrosine with [$^{18}$F]fluorine in HF gave predominantly 3-[$^{18}$F]fluoro-L-α-methyltyrosine (**9.35**) (Scheme 9.9). [$^{18}$F]Acetylhypofluorite gave a mixture of 2- and

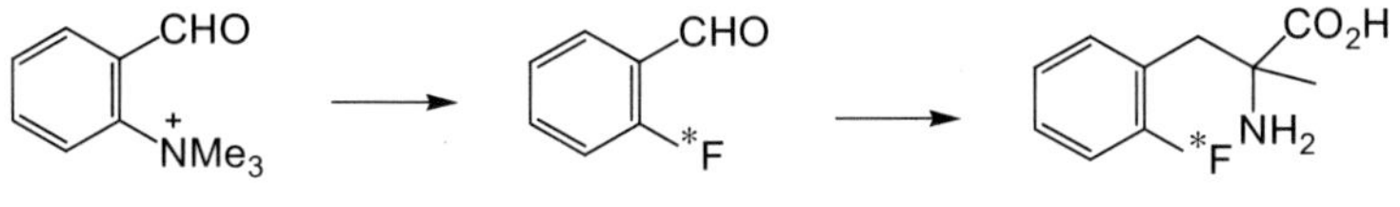

**Scheme 9.7** The synthesis of [$^{18}$F]-α-methylphenylalanine.

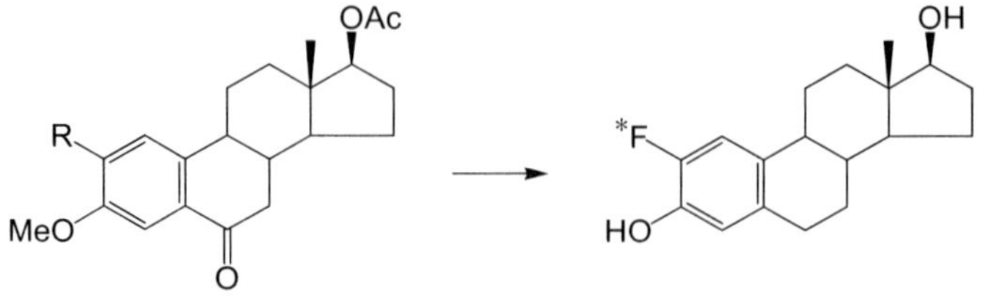

9.32   R = H

9.33   R = $\overset{+}{N}Me_3$

9.34

**Scheme 9.8**   The synthesis of 2-[$^{18}$F]fluoroestradiol.

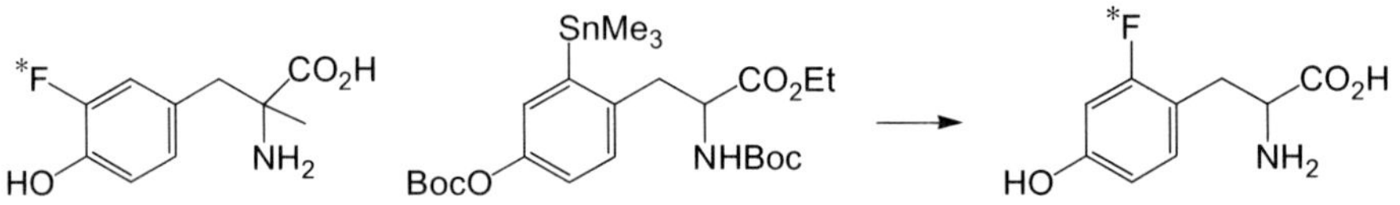

9.35                              9.36                              9.37

**Scheme 9.9**   The synthesis of 2-[$^{18}$F]fluorotyrosine.

3-fluoro compounds whereas fluorination in the presence of boron tri-fluoride gave the 3,5-difluoro compound. Fluorodestannylation reactions have been carried out with [$^{18}$F]selectfluor bis(triflate) and have also been used to label the aromatic ring of tyrosine and L-DOPA. Thus fluorination of the Boc-protected stannane **9.36** and treatment with HBr to remove the protecting groups gave 2-[$^{18}$F]fluoro-L-tyrosine (**9.37**).

Enzymatic fluorination is a rare and relatively unexplored area of biotransformation. However, the isolation of a fluorinase enzyme from the bacterium *Streptomyces cattleya* has led to an application to the labelling of a sugar with fluorine-18. Incubation of *S*-adenosyl-L-me-thionine with *S. cattleya* fluorinase gave 5′-fluoro-2-deoxyadenosine. By coupling this enzyme with an L-amino acid oxidase to remove the L-methionine that was formed and with a purine nucleotide phosphorylase together with a phytase, it was possible to obtain 5′-[$^{18}$F]fluoro-2-de-oxyribose (**9.38**) within a time-scale that brought the method into the realm of a potential fluorination technique for use in PET studies.

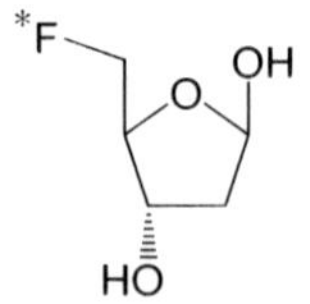

9.38 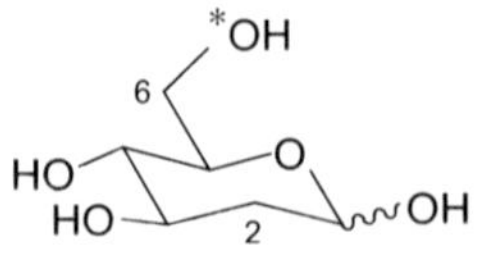                              9.39

## 9.5  NITROGEN-13 IN PET

Nitrogen-13 is a positron-emitting isotope of nitrogen with a short half-life (9.96 min). It has been prepared as [$^{13}$N]ammonia by the proton bombardment of the oxygen in water using the nuclear reaction $^{16}$O(p, $\alpha$)$^{13}$N or by the deuteron bombardment of methane by the reaction $^{12}$C(d, n)$^{13}$N. The [$^{13}$N]ammonia has been used in the enzymatic synthesis of L-[$^{13}$N]glutamic acid, L-[$^{13}$N)glutamine, L-[$^{13}$N]alanine and L-[$^{13}$N]aspartic acid using immobilized enzymes such as glutamate dehydrogenase, glutamate-pyruvate transaminase, glutamine synthetase and aspartase with the corresponding $\alpha$-keto acids or fumaric acid as substrates. The labelled products were used in the assessment of myocardial metabolism. [$^{13}$N]Ammonia has also been oxidized to [$^{13}$N]nitrous acid and used to prepare [$^{13}$N]-*N*-nitrosoureas.

## 9.6  OXYGEN-15 IN PET

Although oxygen-15 is a positron-emitting isotope, it has had only limited application because of its very short half-life (2 min). One application involved the oxidation of tributylborane to form [1-$^{15}$O]butanol, which was used to monitor blood flow. 6-[$^{15}$O]-2-Deoxy-D-glucose (**9.39**) has been prepared by oxygenation of a radical prepared from 2, 6-dideoxy-6-iodo-D-glucose with tri-*n*-butylstannane and azobisisobutyronitrile. The peroxide was reduced with triphenylphosphine. This oxygenation methodology has wider applicability in the context of preparing oxygen-15-labelled compounds.

## 9.7  ISOTOPES OF BROMINE

Apart from the common stable isotopes bromine-79 and bromine-81, there are a number of radioactive isotopes of bromine. Among these are $^{75}$Br ($t_{1/2}$ 97 min) and $^{76}$Br ($t_{1/2}$16.2 h). The latter, with a longer half-life, has had limited use in PET imaging. They are produced by bombardment of arsenic or selenium alloys. The maximum positron energy of $^{76}$Br is fairly high (3.9 MeV), producing rather diffuse images. Electrophilic bromine can be produced from sodium [$^{76}$Br]bromide by oxidation with hydrogen peroxide and *N*-chlorosuccinimide in acetic acid. This has been used in destannylation reactions in, for example, the preparation of [$^{76}$Br]-5-bromo-2-fluorouracil (**9.41**) from the corresponding 5-trimethylstannyl derivative **9.40** (Scheme 9.10). This analogue of a nucleic acid base is incorporated into DNA and has been used for tumour imaging. Copper-catalysed nucleophilic exchange reactions can also be used to introduce bromine-76. For example,

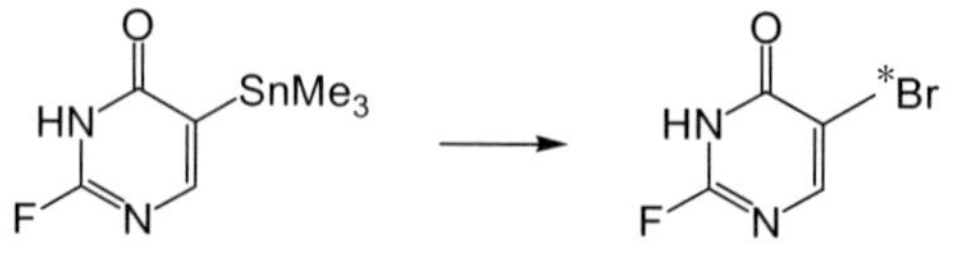

**Scheme 9.10**   The synthesis of 5-[$^{76}$Br]bromo-2-fluorouracil.

[$^{76}$Br]bromoepibatidine (**9.42**) has been produced by exchange of a re-active halogen on the pyridine ring using copper sulfate as the catalyst. This labelled derivative has potential application in the study of neu-rodegenerative diseases such as Alzheimer's disease.

  Bromine-82 is obtained by irradiating natural bromine with neutrons in a nuclear reactor. It is formed together with bromine-80, which has a significantly shorter half-life. Bromine-82 decays with a half-life of 35.4 h with the emission of a β-particle and γ-radiation. This means that it is suitable for use in single photon emission computed tomography (SPECT). Similar chemistry involving the formation of electrophilic bromine by oxidation of the bromide ion has been used to introduce bromine-82 on to the pyrimidine ring of uracil. It has also been used in the electrophilic aromatic substitution of activated rings as in the estrogens.

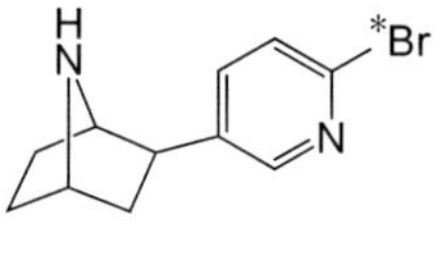

**9.42** 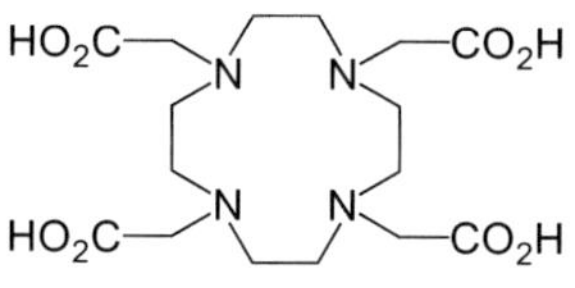                          **9.43** 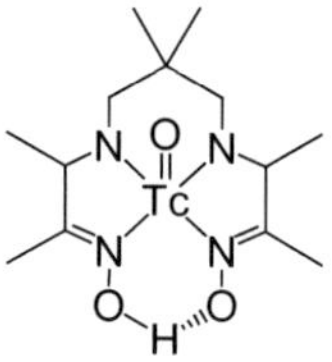                          **9.44**

## 9.8   POSITRON-EMITTING METALLIC ISOTOPES IN PET

The scope of PET has been widened to include a number of β$^+$-emitting isotopes of metals. These include $^{64}$Cu ($t_{1/2}$ 13 h), $^{66}$Ga ($t_{1/2}$ 9.5 h), $^{68}$Ga ($t_{1/2}$ 68 min) and $^{86}$Y ($t_{1/2}$ 15 h). The strategy is to attach a complexing agent such as 1,4,7,10-tetraazacyclodecane-1,4,7,10-tetraacetic acid (DOTA) (**9.43**) to a biomolecule such as the peptides bombasin, integrin or somatostatin. The latter selectively bind to specific, often cancerous, cells while the complex carries the metal. Sometimes a spacer unit is introduced between the DOTA and the peptide to prevent the complex affecting the binding properties of the peptide. These can have value in facilitating the detection of metastases of primary tumours. This methodology can also be used to bind selectively isotopes such as $^{67}$Cu which are $\beta^-$ emitters and are used as radiotherapeutic agents.

## 9.9 SINGLE PHOTON EMISSION COMPUTED TOMOGRAPHY

A number of isotopes that decay with the emission of $\gamma$-radiation have a half-life which makes them suitable for use in single photon emission computed tomography (SPECT). Their chemistry allows them to be attached either directly or as part of a complex to biomolecules that target particular organs. These isotopes include technetium-99m and iodine-123.

## 9.10 TECHNETIUM-99M

The element technetium does not occur naturally but is produced artificially. The metastable isotope of technetium-99 is produced by the decay of molybdenum-99, which in turn is usually obtained as a fission product of uranium-235. The molybdenum-99 can also be formed from the stable isotope molybdenum-98 by neutron capture. The isotope $^{99m}$Tc has a half-life of 6 h and is a $\gamma$-ray emitter. The energy of this emission (141 keV) makes it suitable for SPECT. The technetium-99m is obtained in a very dilute aqueous solution as $[TcO_4]^-$ by elution of a column containing its parent as $[^{99}MoO_4]^-$. A number of its applications have involved the formation of complexes that are designed to keep it in solution and which are used to image blood circulation. Because of its radioactivity and scarcity, much of the organic chemistry involved in the synthesis of technetium complexes has been explored in the first instance by using rhenium, which occurs in the same group of the Periodic Table. Complexes such as the hexamethylpropylenediamine oxime (**9.44**) (Ceretec®) and ethylenedicysteine diethyl ester (**9.45**) (Neurolite®) are sufficiently lipophilic to cross the blood–brain barrier. They have been used to monitor blood flow in the brain and to reveal regions which have been damaged by a stroke. Other complexes have been used to monitor heart, kidney and liver function and bone structure. A further group of complexes are hybrids of a unit designed to bind selectively to specific receptors and a second unit to chelate the technetium. Examples are the steroid **9.46**, which is structurally related to mifepristone, and the cocaine analogue **9.47** (Trodat®), which target the progesterone and dopamine receptors, respectively. Technetium-99m complexes linked to carbohydrates exemplified by **9.48** have been synthesized as potential

SPECT imaging agents. This compound exemplifies the use of a $[^{99m}Tc(CO)_3]^+$ core and the strategy of linking a complexing agent to a biomolecule.

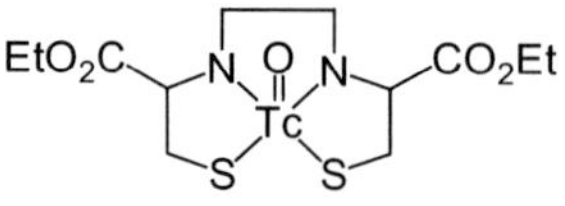

**9.45**

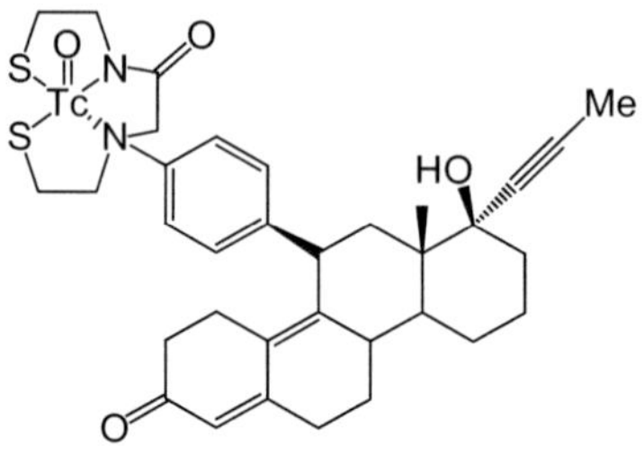

**9.46**

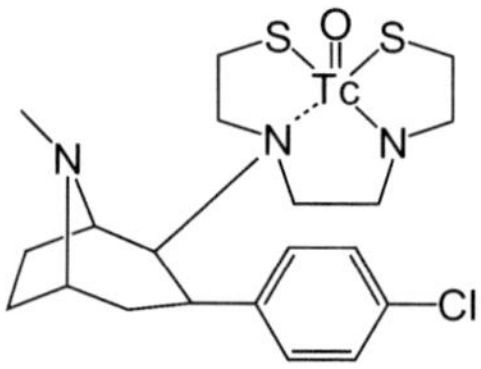

**9.47**

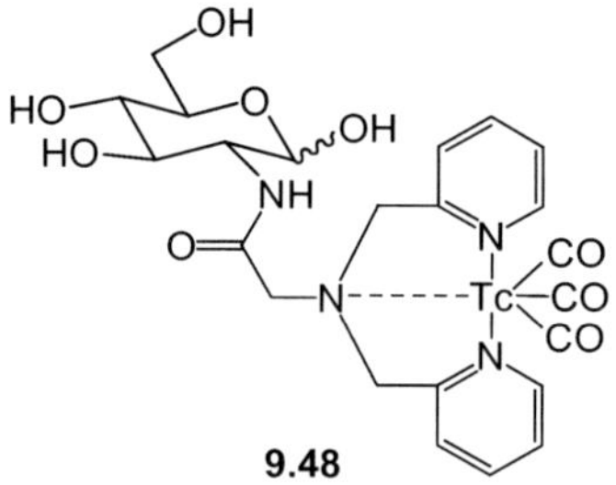

**9.48**

Attempts have also been made to synthesize compounds in which the technetium (or rhenium) forms an integral part of a structure which possesses an overall formal resemblance to a hormone such as a steroid. In designing these compounds, consideration has to be given not just to the radioisotopic half-life but also to the biological half-life of the technetium complex.

## 9.11  IODINE-123

Iodine-123 is produced by the bombardment of xenon-124 with protons in a sequence of nuclear reactions:

$$^{124}Xe(p, 2n)^{123}Cs \rightarrow {}^{123}Xe \rightarrow {}^{123}I$$
$$^{124}Xe(p, pn)^{123}Xe \rightarrow {}^{123}I$$

The xenon-123 decays with a half-life of 2.08 h. Iodine-123 has a half-life of 13.2 h and decays by electron capture with the emission of a photon. The main use of iodine-123 is for SPECT. The iodine-123 is produced as sodium [$^{123}$I]iodide in alkaline solution. It has been used to prepare iodinated fatty acids for the study of the their α-oxidation in myocardial cells and in the preparation of 2β-carbomethoxy-3β-(4'-[$^{123}$I]iodophenyl)tropanes (**9.8**) for imaging the dopamine transporter. Alterations in the function of the dopamine transporter have been associated with Parkinson's disease. The $^{123}$I was introduced on to the aromatic ring by displacement of the trimethylstannane derivative with the iodonium ion prepared by oxidation of the sodium [$^{123}$I]iodide with peracetic acid.

In this chapter, we have seen that the rapidly developing use of radiopharmaceuticals in medical imaging involves the application of both organic and inorganic chemistry. The synthetic chemistry involves not just enhancing the yield in key steps but also developing a sequence of reactions which can be completed within a very short period on a micromolar scale.

# Conclusion

In this book, we have seen the variety of ingenious methods that have been used to introduce different isotopic labels into organic compounds for applications that range from mechanistic organic chemistry through biochemistry to medical imaging and forensic analysis. The last 75 years have seen many developments in the application of synthetic methods to these isotopic labelling problems.

The conventional approach to the design of a synthesis of an unlabelled molecule is based on a retrosynthetic analysis of the target structure. This leads to dissections which are based on identifying relationships between functional groups and strategic bonds which need to be modified for the synthesis of labelled molecules. Many of the examples in this book show that in designing an efficient and economic synthesis of a labelled compound, there are other factors that need to be taken into account. First, there are a number of simple building blocks such as the cyanide ion, methyl iodide, acetic acid, diethyl malonate, ethyl acetoacetate and glycine which are readily available in a labelled form. Hence an important criterion in evaluating a dissection is to consider the potential building blocks which might be used and whether these can be introduced at a late stage in the synthesis. Indeed, it is important to inspect the target structure for 'marker' units which might indicate their use. Second, there are a number of structural units that are easily labelled *in situ* by, for example, exchange or hydrogenolysis reactions. Inspection of a structure for the presence of these units, or for simple dissections which might reveal them, can provide a useful labelling strategy. Third, there are many recently discovered reactions, particularly organometallic reactions and palladium-catalysed coupling reactions, which have considerable potential for labelling compounds.

The Organic Chemistry of Isotopic Labelling
By James R. Hanson
© James R. Hanson 2011
Published by the Royal Society of Chemistry, www.rsc.org

Although a number of these have been exploited in the context of the rapid introduction of a label for positron emission tomography, some of their wider applications in labelling compounds remain to be developed. Indeed, many strategies which have been developed in the context of one isotope have the potential for use with other isotopes. Careful consideration of the mechanisms of these newly discovered reactions can suggest simple ways in which they may be used for labelling organic compounds. Thus reactions which are terminated by protonation may be used to incorporate deuterium or tritium from deuterium oxide or tritiated water. Selective *ortho* exchange reactions also have wide potential, particularly when it is recognized that a significant number of pharmaceutically interesting compounds possess an aromatic ring bearing a substituent which can be used, or modified, to direct such a reaction.

The labelling of complex chiral natural products and related pharmaceuticals provides another series of problems. In the majority of cases, total synthesis has an impracticable time-scale and is uneconomic in terms of the loss of label. Simple degradation and re-synthesis can provide the answer. These strategies, and the use of another natural product as a chiral template in order to construct an enantiospecifically labelled target molecule, require a knowledge of the key reactions that have been used in the past in the degradation of natural products.

The biosynthetic and particularly the stereochemical consequences of many enzyme reactions are known. Biotransformations can provide the opportunity for labelling complex molecules, often from simple sources of the label and in a chiral manner. Consequently, the combination of chemical and enzymatic methods, exemplified in the synthesis of labelled amino acids, can provide an attractive and economic solution to a labelling strategy.

The small scale on which many synthetic labelling sequences are carried out makes them particularly amenable to various rate-enhancement techniques such as microwave radiation and the use of ionic liquids.

The consequences of these developments in synthetic strategies have implications for the planning of the synthesis of a known labelled compound. It is sensible when repeating a known strategy to consider what was available at the time the synthesis was originally developed and whether novel means of carrying out particular transformations might be more appropriate today.

The increased sensitivity of analytical methods, particularly in the measurement of isotope ratios, NMR detection and in the use of positron emission tomography, have not only widened the scope of problems that can be tackled but also altered the constraints that the synthetic strategies have to fulfil. In particular, the sensitive mass spectrometric measurement of isotope ratios and the influence of one isotope on another in the NMR spectrum, in terms of both chemical shift and coupling pattern, bring a further range of problems concerning the origin of compounds and the integrity of biosynthetic units within the scope of isotopic methods. These will require the synthesis of appropriately labelled compounds. Apart from the conventional problems of mechanistic organic chemistry and metabolic studies, there are now many more problems of a medical, environmental and ecological nature that are potentially susceptible to isotopic study, providing new areas for future investigations.

# Further Reading

There are many specialist books and reviews dealing with specific aspects of the chemistry of isotopic labelling and the handling of radioisotopes. The following selection should provide an entry into the primary literature. The *Journal of Labelled Compounds and Radiopharmaceuticals* contains not only research papers but also reviews and recently a classified 'current awareness' bibliography. The regular volumes of the *Synthesis and Applications of Isotopes and Isotopically Labelled Compounds* (Vol. 8, 2004, Wiley, Chichester) contain papers from the international isotope meetings. General information can be obtained from the textbook *An Introduction to Radiochemistry*, D.J. Malcolm-Lawes (Macmillan, London, 1979). The textbook *Radioisotope Laboratory Techniques*, R.A. Faires and B.H. Parks (Butterworths, London, 1973) contains useful information on techniques, while more recent guidance covering a range of applications is available from the technical information produced by commercial suppliers, *e.g.* GE Healthcare (Amersham Radiochemicals). The booklets produced by the former Radiochemical Centre contain very helpful information and references to the primary literature and to specialist books on, for example, liquid scintillation counting. The references to the primary literature have been selected to exemplify different methodologies.

## CHAPTER 1

The *Annual Reports of the Chemical Society* between 1904 and 1943 contain regular chapters on 'Radioactivity' by Frederick Soddy and

The Organic Chemistry of Isotopic Labelling
By James R. Hanson

subsequently by others on 'Radioactivity and sub-atomic phenomena'. These cover the early developments of the subject. The chapters on 'Isotope chemistry' by A.R. Ubbelohde (*Annu. Rep. Chem. Soc.*, 1949, **46**, 9) and 'The function of small molecules in biosynthesis' by R. Bentley (*Annu. Rep. Chem. Soc.*, 1948, **45**, 239) review early chemical and biochemical applications. Some other relevant articles include 'Thermodynamic properties of isotopic substances', H.C. Urey, *J. Chem. Soc.*, 1947, 562.

The preparation of radioactive tracers, F.A. Paneth, *Q. Rev. Chem. Soc.*, 1948, **2**, 93.

Isotopic tracer techniques, H.R.V. Arnstein and R. Bentley, *Q. Rev. Chem. Soc.*, 1950, **4**, 172.

Early history of carbon-14, M.D. Kamen, *J. Chem. Educ.*, 1963, **40**, 234.

NMR and the Periodic Table, R.K. Harris, *Chem. Soc. Rev.*, 1976, **5**, 1.

Carbon-13 nuclear magnetic resonance in biosynthetic studies, T.J. Simpson, *Chem. Soc. Rev.*, 1975, **4**, 4.

Some new NMR methods for tracing the fate of hydrogen in biosynthesis, M.J. Garson and J. Staunton, *Chem. Soc. Rev.*, 1979, **8**, 539.

Applications of multinuclear NMR to structural and biosynthetic studies of polyketide microbial metabolites, T.J. Simpson, *Chem. Soc. Rev.*, 1987, **16**, 123.

Isotope fractionation of organic compounds in chromatography, C.N. Filer, *J. Label. Compd. Radiopharm.*, 1999, **42**, 169.

Isotopic labelling in the study of organic and organometallic mechanism and structure: an account, G.C. Lloyd-Jones and M. Paz Munoz, *J. Label Compd. Radiopharm.*, 2007, **50**, 1072.

## CHAPTER 2

*Carbon-14 Compounds*, J.R. Catch, Butterworths, London, 1961.

*Syntheses with Stable Isotopes of Carbon, Nitrogen and Oxygen*, D.G. Ott, Wiley, New York, 1981.

*Preparation of Compounds Labelled with Tritium and Carbon-14*, R. Voges, J.R. Heys and T. Moenius, Wiley, Chichester, 2009.

Synthesis of isotopically labelled compounds, S.L. Thomas and H.S. Turner, *Q. Rev. Chem. Soc.*, 1953, **7**, 407.

The synthesis of radiolabelled compounds via organometallic intermediates, G.W. Kabalka and R.S. Varma, *Tetrahedron*, 1989, **45**, 6601.

Chemical preparation of asymmetrically labelled citric acid, P.E. Wilcox, C. Heidelberger and V.R. Potter, *J. Am. Chem. Soc.*, 1950, **72**, 5019.

The alkaline rearrangement of α-haloketones. The mechanism of the Favorskii rearrangement, R.B. Lotfield, *J. Am. Chem. Soc.*, 1951, **73**, 4707.

[6-$^{14}$C]-D-Glucose and [6-$^{14}$C]-D-glucuronolactone, J. Sowden, *J. Am. Chem. Soc.*, 1952, **74**, 4377.

(–)-Kaurene as a precursor of gibberellic acid, B.E. Cross, R.H.B. Galt and J.R. Hanson, *J. Chem. Soc.*, 1964, 295.

Preparation of [1, 2-$^{13}$C$_2$]abscisic acid for use as a stable and pure internal standard, T. Asami, K. Sekimata, J.M. Wang, K. Yoneyama, Y. Takeuchi and S. Yoshida, *J. Chem. Res.*, 1999, 658.

The synthesis of singly and doubly $^{13}$C-labelled mevalonolactone, J.A. Lawson, W.T. Colwell, J.I. DeGraw, R.H. Peters, R.L. Dehn and M. Tanabe, *Synthesis*, 1975, 729.

50 Years of the synthesis of labelled mevalonolactone, J.R. Hanson, *J. Chem. Res.*, 2008, 241.

Biosynthesis of mycophenolic acid, L. Canonica, W. Kroszczynski, B.M. Ranzi, B. Rindone, A. Santaniella and C. Scolastico, *J. Chem. Soc., Perkin Trans. 1*, 1972, 2639.

A convenient synthesis of 4-hydroxy[1-$^{13}$C]benzoic acid and related ring-labelled phenolic compounds, J. Beyer, S. Lang-Fugmann, A. Muhlbauer and W. Steglich, *Synthesis*, 1998, 1047.

Chemical synthesis of $^{13}$C and $^{15}$N labelled nucleosides, I.M. Lagoja and F. Herdewijn, *Synthesis*, 2002, 301.

Synthesis of $^{14}$C-labelled guanine, adenine, 8-azaguanine and 8-azaadenine, E.L. Bennett, *J. Am. Chem. Soc.*, 1952, **74**, 2420.

Synthesis of some purines and pyrimidines labelled with $^{14}$C, L.J. Bennett, *J. Am. Chem. Soc.*, 1952, **74**, 2432.

Methods for the synthesis of carbon-13 labelled acids and esters, A.C. Jordan, L.C. Oxford, J.R. Harding, Y. O'Connell, T.J. Simpson and C. Willis, *J. Label. Compd. Radiopharm.*, 2007, **50**, 338.

Synthesis of all-*trans* beta-carotene retinoids and derivatives labelled with $^{14}$C, E. Azim, P. Auzeloux, J.-C. Maurizis, V. Braesco, P. Grolier, A. Veyre and J.C. Madelmont, *J. Label. Compd. Radiopharm.*, 1996, **38**, 441.

Fast and efficient synthesis of $^{14}$C labelled benzonitriles and their corresponding acids, S.C. Schou, *J. Label. Compd. Radiopharm.*, 2009, **52**, 173.

Synthesis of a series of carbon-14 labelled 4-aminoquinazolines and quinazolin-4(3H)-ones, N.Saemian, O.K.Arjomandi and G.Shirvani, *J. Label. Compd. Radiopharm.*, 2009, **52**, 453.

## CHAPTER 3

*Tritium and its Compounds*, E.A. Evans, Butterworths, London, 1966.

*Deuterium Labelling in Organic Chemistry*, A.F. Thomas, Appleton-Century-Crofts, New York, 1981.

Introduction of deuterium into the steroid system, L. Tokes and L.J. Throop, in *Organic Reactions in Steroid Chemistry,* ed. J. Fried and J.A. Edwards, Van Nostrand Reinhold, New York, 1972.

Isotope effects in organic chemistry and biochemistry, H. Simon and D. Palm, *Angew. Chem. Int. Ed. Engl.*, 1966, **5**, 920.

Isotopes and organic reaction mechanisms, C.J. Collins, *Adv. Phys. Org. Chem.*, 1964, **2**, 3.

Tritium chemistry: history, current status and future developments, W.J.S. Lockley, *J. Label. Compd. Radiopharm.*, 2007, **50**, 256.

Synthetic tritium labelling, reagents and methodologies, M. Saljoughan, *Synthesis*, 2002, 1781.

The synthesis of alcohols-OD, L. Verbit, *Synthesis*, 1972, 254.

Deuterium and tritium exchange reactions of phenols and the synthesis of labelled 3,4-dihydroxyphenylalanines, G.W. Kirby and L. Ogunkoya, *J. Chem. Soc. C*, 1965, 6914.

Aldehydes from dihydro-1,3-oxazines, synthesis of aliphatic aldehydes and their C-1 deuteriated derivatives, A.I. Meyers, A. Nabeya, H. Adickes and I.R. Politzer, *J. Am. Chem. Soc.*, 1969, **91**, 763.

Conversion of carboxylic acids into aldehyde and their C-1 or C-2 deuteriated derivatives, J. Cymerman Craig, N.N. Ekwuribe, C.C. Fu and K.A.M. Walker, *Synthesis*, 1981, 303.

Investigations into the effectiveness of deuterium as a 'protecting group' for C–H bonds in radical reactions involving hydrogen atom transfer, M.E. Wood, S. Bissiriou, C. Lowe and K.M. Windeate, *Org. Biomol. Chem.*, 2008, **6**, 3048.

Ligand effects upon deuterium exchange in arenes mediated by $[Ir(PR_3)_2(cod)]^+BF_4^-$, G.J. Ellames, J.S. Gibson, J.M. Herbert, W.J. Kerr and A.H. McNeill, *J. Label. Compd. Radiopharm.*, 2004, **47**, 1.

Tritium and deuterium labelling studies of alkali metal borohydrides and their application to simple reductions, C. Than, H. Morimoto, H. Andres and P.G. Williams, *J. Label. Compd. Radiopharm.*, 1996, **38**, 693.

Preparation of site specifically deuterated 7,12-dimethylbenz[*a*]an-
thracene derivatives:mechanism of hydrogenolysis of aryl halides
with lithium aluminium hydride, S.R. Adapa, Y.M. Sheikj, R.W.
Hart and D.T. Witiak, *J. Org. Chem.*, 1980, **45**, 3343.
Microwave enhanced hydrogenation reactions using solid hydrogen,
deuterium and tritium donors, M. Al-Qahtani, N. Cleator, T.N.
Danks, R.N. Garman, J.R. Jones, S. Stefaniak, A.D. Morgan and
A.J. Simmonds, *J. Chem. Res. (S)*, 1998, 400.
Microwave enhanced decarboxylations of aromatic carboxylic acids,
L.B. Frederiksen, T.H. Grobosch, J.R. Jones, S.-Y. Lu and C.C.
Zhao, *J. Chem. Res.*, 2000, 42.
Some deuterium labelling experiments, J.R. Hanson, *J. Chem. Educ.*,
1982, **59**, 342.
Convenient preparations of deuterium-substituted furan thiophen- and
*N*-methylpyrrole-2-carboxylic acids and of tetradeuteriofuran, D.J.
Chadwick, J. Chambers, G.D. Meakins and R.L. Snowden, *J. Chem.
Soc., Perkin Trans. 1*, 1973, 201.
Diazomethane-d$_2$, J.R. Cambell, *Chem. Ind. (London)*, 1972, 540.
Regioselective deuteriation of aromatic and αβ-unsaturated carboxylic
acids via rhodium(III) chloride catalysed exchange with deuterium
oxide, W.J.S. Lockley, *Tetrahedron Lett.*, 1982, **23**, 3819.
Parallel chemistry investigations of *ortho*-directed hydrogen isotope
exchange between substituted aromatics and isotopic water, L.P.
Kingston, W.J.S. Lockley, A.N. Mather, E. Spink, S.P. Thompson
and D.K. Wilkinson, *Tetrahedron Lett.*, 2000, **41**, 2705.
Reactions of phosphines with acetylenes, synthesis of 1,2-dideuteriated
olefins, E.M. Richards, J.C. Tebby, R.S. Ward and D.H. Williams, *J.
Chem. Soc. C*, 1969, 1542.
Synthesis of 1-deuterioalkyl halides from *S*-alkylsulfonium salts
and heavy water, H. Dorn, *Angew. Chem. Int. Ed. Engl.*, 1967, **6**,
371.
Preparation of all the possible ring-deuteriated benzoic acids by reductive dehalogenation of the corresponding halogenobenzoic acids
with Raney alloys in an alkaline deuterium oxide solution, M.
Tashiro, K. Nakayama and G. Fukata, *J. Chem. Soc., Perkin Trans.
1*, 1983, 2315.
One step synthesis of deuterium or tritium labelled imines and aldazines
under mild conditions, D.H.R. Barton, E. Doris and F. Taran, *J.
Label. Compd. Radiopharm.*, 1998, **41**, 871.
Synthesis and characterization of [*phenyl*-³H]clonidine hydrochloride at
high specific activity, C.N. Filer and D.G. Ahern, *J. Label. Compd.
Radiopharm.*, 2001, **44**, 323.

General method of obtaining deuterium labelled heterocyclic compounds using neutral $D_2O$ with heterogeneous Pd/C, H. Esaki, N. Ito, S. Sakai, T. Maegawa, Y. Monguchi and H. Sajiki, *Tetrahedron*, 2006, **62**, 10954.

An improved method for preparing tritium labelled fluoxetine, R.S.P. Hsi and W.T. Stolle, *J. Label. Compd. Radiopharm.*, 1996, **38**, 1148.

A new practical labelling procedure using sodium borotritide and tetrakis(triphenylphosphine)palladium(0), T. Nagasaki, K .Sakai, M. Segawa, Y. Katsuyama, N. Haga, M. Koike, K. Kawada and S. Takechi, *J. Label. Compd. Radiopharm.*, 2001, **44**, 993.

Synthesis of deuterium-labelled chlorhexidine, M. Moser, T. Hudlicky, S. Sadeghi and E. Sternin, *J. Label. Compd. Radiopharm.*, 2007, **50**, 671.

New amine-stabilized deuteriated borane–tetrahydrofuran complex ($BD_3$–THF); convenient reagent for deuterium incorporations, R.C. Todd, M.M. Hossain, K.V. Tosyula, P. Gao, J. Kuo and C.T. Tan, *Tetrahedron Lett.*, 2007, **48**, 2335.

Easy preparative scale synthesis of labelled xanthines, caffeine, theophylline and theobromine, F. Balssa and Y. Bonnaire, *J. Label. Compd. Radiopharm.*, 2007, **50**, 38.

**CHAPTER 4**

Some applications of tritium labelling for the exploration of biochemical mechanisms, A.R. Battersby, *Acc. Chem. Res.*, 1972, **5**, 148.

Fragmentation and hydrogen transfer reactions of a typical 3-keto steroid, 5α-androstan-3-one, R.H. Shapiro, D.H. Williams, H. Budzikiewicz and C. Djerassi, *J. Am. Chem. Soc.*, 1964, **86**, 2837.

One-pot deuteriation and reduction of ketones in the synthesis of $[16,16,17-^2H_3]$-epitestosterone, H. Chodounska, A. Kasal, D. Saman and K. Ubik, *Collect. Czech. Chem. Commun.*, 1996, **61**, 1037.

Chemistry of mevalonic acid, J.W. Cornforth and R.H. Cornforth, in *Natural Substances Formed Biologically from Mevalonic Acid*, ed. T.W. Goodwin, Academic Press, London, 1970, p. 5.

Exploration of enzyme mechanisms by asymmetric labelling, J.W. Cornforth, *Q. Rev. Chem. Soc.*, 1969, **23**, 125.

The chiral methyl group, its biochemical significance, J.W. Cornforth, *Chem. Br.*, 1970, **6**, 431.

A new synthesis of chiral acetic acid, C.A. Townsend, T. Scholl and D. Arigoni, *Chem. Commun.*, 1975, 921.

Highly stereoselective synthesis of a chiral methyl group by a facially controlled sigmatropic [1,5]-hydrogen shift, C. Dehnardt, M.

McDonald, S. Lee, H.G. Floss and J. Mulzer, *J. Am. Chem. Soc.*, 1999, **121**, 10848.

Incorporation of deuterium-labelled analogs of isopentenyl diphosphate for the elucidation of the stereochemistry of rubber biosynthesis, A.A. Scholte and J.C. Vederas, *Org. Biomol. Chem.*, 2006, **4**, 730.

Enzyme-assisted preparation of isotope-labelled 1-deoxy-D-xylulose 5-phosphate, S. Hecht, K. Kis, W. Eisenreich, S. Amslinger, J. Wungsintaweekui, S. Herz, F. Rohdich and A. Bacher, *J. Org. Chem.*, 2001, **66**, 3948.

The synthesis of isotopically labelled *N*-acetylcysteamine thioesters utilising a baker's yeast reduction in $D_2O$, M.P. Dillon, M.A. Hayes, T.J. Simpson and J.B. Sweeney, *Bioorg. Med. Chem. Lett.*, 1991, **1**, 223.

Stereochemistry of enzymic hydrogen transfer to pyridine nucleotides, J.W. Cornforth, G. Ryback, G. Popjak, C. Donninger and G. Schroepfer, *Biochem. Biophys. Res. Commun.*, 1962, **8**, 371.

Stereochemistry of the succinic dehydrogenase reaction, T.T. Chen and H. van Milligan, *J. Am. Chem. Soc.*, 1960, **82**, 4115.

The stereospecific conversion of stearic acid to oleic acid, G.J. Schroepfer and K. Bloch, *J. Biol. Chem.*, 1965, **240**, 54.

## CHAPTER 5

Synthesis of L-serine stereospecifically labelled at C-3 with deuterium, D. Gani and D.W. Young, *J. Chem. Soc., Perkin Trans. 1*, 1981, 2393.

Stereospecific synthesis of $(2S,4R)$-[5,5,5-$^2H_3$]-leucine, R.A. August, J.A. Khan, C.M. Moody and D.W. Young, *Tetrahedron Lett.*, 1992, **33**, 4617.

Versatile synthesis of stereospecifically labelled D-amino acids via labelled aziridines, B.S. Axelsson, K.J. O'Toole, P.A. Spencer and D.W. Young, *J. Chem. Soc., Perkin Trans. 1*, 1994, 807.

Synthesis of $(2S, 3S)$-[3-$^2H_1$]-4-methyleneglutamic acid and $(2S, 3R)$-[2,3,-$^2H_2$)-4-methyleneglutamic acid, P. Dieterich and D.W. Young, *Org. Biomol. Chem.*, 2006, **4**, 1492.

Two separate and distinct syntheses of stereospecifically deuteriated samples of (2S)-proline, P. Barraclough, P. Dieterich, C.A. Spray and D.W. Young, *Org. Biomol. Chem.*, 2006, **4**, 1483.

Three approaches to the synthesis of L-leucine selectively labelled with carbon-13 or deuterium in either diastereotopic methyl group, M.D. Fletcher, J.R. Harding, R.A. Hughes, N.M. Kelly, H. Schmalz, A. Sutherland and C.L. Willis, *J. Chem. Soc., Perkin Trans. 1*, 2000, 43.

A short versatile chemical synthesis of ʟ- and ᴅ-amino acids stereo-
selectively labelled solely in the β-position, K. Lowpetch and D.W.
Young, *Org. Biomol. Chem.*, 2005, **3**, 3348.
Synthesis of chirally labelled cysteines and the steric origin of C(5) in
penicillin biosynthesis, D.J. Morecombe and D.W. Young, *Chem.
Commun.*, 1975, 198.
Synthesis of [6-$^{13}$C]lysine, A. Sutherland and C.L. Willis, *J. Label.
Compd. Radiopharm.*, 1995, **38**, 95.


## CHAPTER 6

Aromatic hydroxylation of lidocaine and mepivacaine in rats and
humans. J. Thomas and P. Meffin, *J. Med. Chem.*, 1972, **15**, 1046.
*In vivo* metabolite condensations. Formation of *N'*-ethyl-2-methyl-*N$^3$*-
(2,6-dimethylphenyl)-4-imidazoidinone from the reaction of a metabolite of alcohol with a metabolite of lidocaine, S.D. Nelson, G.D.
Breck and W.F. Trager, *J. Med. Chem.*, 1973, **16**, 1106.
Primary and β-secondary deuterium isotope effects in *N*-deethylation
reactions, S.D. Nelson, L.R. Pohl and W.F. Trager, *J. Med. Chem.*,
1975, **18**, 1062.
An efficient laboratory synthesis of α-deuteriated profens, G.S. Coum-
barides, M. Dingjan, J. Eames, A. Flinn and J. Northen, *J. Label.
Compd. Radiopharm.*, 2006, **49**, 903.
A novel asymmetric synthesis of tritium and carbon-14 labelled (*R*)-
ibuprofen, Y. Zhang, Y. Hong, C.C. Huang and D. Belmont, *J.
Label. Compd. Radiopharm.*, 2006, **49**, 237.
Fast and efficient tritium labelling of the non-steroidal anti-inflamma-
tory drugs, naproxen, tolmetin and zomepirac, S.K. Johansen, L.
Sorensen and L. Martiny, *J. Label. Compd. Radiopharm.*, 2005, **48**,
569.
Synthesis of deuterated naproxens, P.M. Gannett, L. Miller, J. Daft, C.
Locuson and T.S. Tracy, *J. Label. Compd. Radiopharm.*, 2007, **50**,
1272.
Synthesis of (RS)-naproxen and its 6-O-demethylated metabolites la-
belled with $^2$H, E.Fontana and S.Venegoni, *J. Label. Compd. Radio-
pharm.*, 2008, **51**, 239.
Synthesis of deuterium labelled phenethylamine derivatives, Y.-Z. Xu
and C. Chen., *J. Label. Compd. Radiopharm.*, 2006, **49**, 1187.
High specific activity (+)-amphetamine and methamphetamine, P.A.
Lamb, C.J. McElhinny, T. Sninski, H. Purdom, F.I. Carrolll and
A.H. Lewin, *J. Label. Compd. Radiopharm.*, 2009, **52**, 457.

Synthesis of [$^2$H$_3$]-labelled sulfamethoxazole and its main urinary metabolites, G. Heinkele and T.E. Murdter, *J. Label. Compd. Radiopharm.*, 2007, **50**, 656.

Preparation of morphine-*N*-methyl-$^{14}$C, H. Rapoport, C.H. Lovell and B.M. Tolbert, *J. Am. Chem. Soc.*, 1951, **75**, 5900.

Morphine-*N*-methyl-$^{14}$C, K.S. Andersen and L.A. Woods. *J. Org. Chem.*, 1959, **24**, 274.

Synthesis of morphine [*N*-methyI-$^{14}$C]-6-β-D-glucuronide, J.R. Ferguson, S.J. Hollis, G.A. Johnston, K.W. Lumbardt and A.V. Stachulski, *J. Label. Compd. Radiopharm.*, 2002, **45**, 107.

Preparation of morphine-6-$^3$H and its isotopic stability in man and in rats, J. Fishman, B. Norton, M.L. Cotter and E.F. Hahn, *J. Med. Chem.*, 1974, **17**, 778.

Protection of the allylic alcohol double bond from catalytic reduction in the preparation of [1-$^3$H]morphine and [1-$^3$H]codeine, H.H. Seltzman, M.J. Roche, C.P. Laudeman, C.D. Wyrick and F.I. Carroll, *J. Label. Compd. Radiopharm.*, 1998, **41**, 811.

Terminal steps in the biosynthesis of the morphine alkaloids, A.R. Battersby, J.A. Martin and E. Brochmann-Hanssen, *J. Chem. Soc. C*, 1967, 1785.

Synthesis of codeine labeled in the 3-methoxy group with carbon-14, F.N.H. Chang, J.F. Oneto, P.P.T. Sah, B.M. Tolbert and H. Rapoport, *J. Org. Chem.*, 1950, **15**, 634.

Synthesis of deuterium-labelled etorphine and dihydroetorphine, Y.J. Chen and C. Chen, *J. Label. Compd. Radiopharm.*, 2007, **51**, 1143.

Application of the Bischler–Napieralski–Pschorr radiosynthesis of (*R*)-(–)-[6α-$^{14}$C]apomorphine, a non-selective D$_1$/D$_2$ dopamine receptor agonist, S.L. Kitson and E. Knagg, *J. Label. Compd. Radiopharm.*, 2006, **49**, 517.

The synthesis of dopaminergic radioligands labelled with tritium and iodine-125, C.N. Filer, *J. Label. Compd. Radiopharm.*, 2007, **50**, 683.

Preparation of (–)-[8,9-$^3$H]apomorphine at high specific activity, C.N. Filer and D.G. Ahern, *J. Org. Chem.*, 1980, **45**, 3918.

Synthesis of deuterium-labelled galanthamine, J. Rouleau and C. Guillou, *J. Label. Compd. Radiopharm.*, 2008, **51**, 236.

Synthesis of deuterium labelled atropine and scopolamine, F. Balssa and Y. Bonnaire, *J. Label. Compd. Radiopharm.*, 2009, **52**, 269.

Synthesis of deuterium labelled cocaine, cocaethylene and metabolites, E.T. Everhart, P. Jacon, J. Mendelson and R.T. Jones, *J. Label. Compd. Radiopharm.*, 1999, **42**, 1265.

Synthesis of *N*-methyl trideuterium-labelled *m*-hydroxybenzoylecgonine as an internal standard for GC/MS analysis, S. Feng and M.A. El-Sohly, *J. Label. Compd. Radiopharm.*, 1999, **42**,1031.
Isotopically labelled tropane alkaloids, S. Patterson and D.A. O'Hagan, *J. Label. Compd. Radiopharm.*, 2002, **45**,191.
Identification and measurement of illicit drugs and their metabolites in urban wastewater by liquid chromatography–tandem mass spectrometry, S. Castiglioni, E. Zuccato, E. Crisci, C. Chiabrando, R. Fanlli and R. Bagnati, *Anal. Chem.*, 2006, **78**, 8421.
Synthesis of deuterium labelled standards of 5-methoxy-*N*,*N*-dimethyl-tryptamine, Y.Z. Xu and C. Chen., *J. Label. Compd. Radiopharm.*, 2006, **49**, 897.

## CHAPTER 7

Chemical synthesis of $^{13}$C and $^{15}$N labelled nucleosides, I.M. Lagoja and P. Herdewijn, *Synthesis*, 2002, 301.
Uses of $^{15}$N as a tracer element in chemical reactions. The mechanism of the Fischer indole synthesis, C.F.H. Allen and C.V. Wilson, *J. Am. Chem. Soc.*, 1943, **65**, 611.
The structure of phenyl azide, K. Clusius and H.R. Weisser, *Helv. Chim. Acta*, 1952, **35**, 1548.
Preparation of optically active lysine labelled with $^{14}$C and $^{15}$N, H.R. V.Arnstein, G.D. Hunter, H.M. Muir and A. Neuberger, *J. Chem. Soc.*, 1952, 1329.
Synthesis of DL-[α-$^{15}$N]tryptophan, G. Tang and X. Tang, *J. Label. Compd. Radiopharm.*, 1999, **42**,199.
Chiral [$^{16}$O,$^{17}$O,$^{18}$O]phosphate esters, G. Lowe, *Acc. Chem. Res.*, 1983, **16**, 244.
Synthesis of [$^{18}$O]phenols from $^{18}$O$_2$, T.E. Walker and M. Goldblatt, *J. Label. Compd. Radiopharm.*, 1984, **21**, 353.
[$^{18}$O]Labelled L-serine, W.F. Karstens and J. Raap, *J. Label. Compd. Radiopharm.*, 1995, **36**, 1077.

## CHAPTER 8

Therapeutic radiopharmaceuticals, W.A. Volkert and T.J. Hoffman, *Chem.Rev.*, 1999, **99**, 2269.
Solid-phase exchange radioiodination of aryl iodides, facilitation by ammonium sulfate, T.L. Mangner, J.-L. Wu and D.M. Wieland, *J. Org. Chem.*, 1982, **47**, 1484.

Radiopharmaceuticals labelled with bromine isotopes, B. Maziere and C. Loc'h, *Appl. Radiat. Isot.*, 1986, **37**, 703.

Radiohalogens for imaging and therapy, M.J. Adam and D.S. Wilbur, *Chem. Soc. Rev.*, 2005, **34**, 153.

Addition reactions of aldehydes and ketones, sulfur isotope effects, W.A. Sheppard and A.N. Bourns, *Can. J. Chem.*, 1954, **32**, 14.

Metabolites associated with organophosphonate C–P bond cleavage. Chemical synthesis and microbial degradation of [$^{32}$P]ethylphosphonic acid, L.Z. Avila, K.M. Draths and J.W. Frost, *Bioorg. Med. Chem. Lett.*, 1991, **1**, 51.

**CHAPTER 9**

Synthesis of $^{11}$C, $^{18}$F, $^{15}$O and $^{13}$N radiolabels for positron emission tomography, P.W. Miller, N.J. Long, R. Vilar and A.D. Gee, *Angew. Chem. Int. Ed.*, 2008, **47**, 8998.

Working against time: rapid radiotracer synthesis and imaging the human brain, J.S. Fowler and A.P. Wolf, *Acc. Chem. Res.*, 1997, **30**, 181.

Microwave enhanced radiochemistry, N. Elander, J.R. Jones, S.-Y. Lu and S. Stone-Elander, *Chem. Soc. Rev.*, 2000, **29**, 239.

The preparation of carbon-11 labelled diprenorphine, S.K. Luthra, V.W. Pike and F. Brady, *Chem. Commun.*, 1985, 1423.

Rapid synthesis of aliphatic amides by reaction of carboxylic acids, Grignard reagents and amines, application to the preparation of [$^{11}$C]amides, C. Aubert, C. Huard-Perrio and M.-C. Lasne, *J. Chem. Soc., Perkin Trans. 1*, 1997, 2837.

Reductive amination of carboxylic acids and [$^{11}$C]magnesium halide carboxylates, C. Perrio-Huard, C. Aubert and M.-C. Lasne, *J. Chem. Soc., Perkin Trans. 1*, 2000, 311.

Biologically active $^{11}$C-labelled amides using palladium-mediated reactions with aryl halides and [$^{11}$C]carbon monoxide, T. Kihlberg and B. Langstrom, *J. Org. Chem.*, 1999, **64**, 9201.

Precursor synthesis and radiolabelling of the dopamine receptor ligand, [$^{11}$C]raclopride from [$^{11}$C]methyl triflate, O. Langer, K. Nagren, F. Dolle, C. Lundkvist, J. Sandell, C.G. Swahn, F. Vaufrey, C. Crouzel, B. Maziere and C. Halldin, *J. Label. Compd. Radiopharm.*, 1999, **42**, 1183.

Aryl triflates and [$^{11}$C/$^{13}$C]carbon monoxide in the synthesis of [$^{11}$C/$^{13}$C]amides, O. Rahman, T. Kihlberg and B. Langstrom, *J. Org. Chem.*, 2003, **68**, 3558.

Fluorine-18 and medical imaging; radiopharmaceuticals for positron emission tomography, D. Le Bars, *J. Fluorine Chem.*, 2006, **127**, 1488.

A solid-phase route to [18]F-labeled tracers, exemplified by the synthesis of [[18]F]2-fluoro-2-deoxy-D-glucose, L.J. Brown, D.R. Bouvet, S. Champion, A.M. Gibson, Y. Hu, A. Jackson, I. Khan, N. Ma, N. Millot, H. Wadsworth and R.C.D. Brown, *Angew. Chem. Int. Ed.*, 2007, **46**, 941.

Synthesis, biodistribution and primate imaging of fluorine-18 labeled 2β-carbo-1′-fluoro-2-propoxy-3β-(4-chlorophenyl)tropanes. Ligands for the imaging of dopamine transporters by positron emission tomography, D. Xing, P. Chen, R. Keil, C.D. Kilts, B. Shi, V.M. Camp, G. Malveaux, T. Ely, M.J. Owens, J. Votaw, M. Davis, J.M. Huffman, R.A.E. BaKay, T. Subramanian, R.L. Watts and M.M. Goodman, *J. Med. Chem.*, 2000, **43**, 639.

Steroids labelled with [18]F for imaging tumors by positron emission tomography, J.A. Katzenellenbogen, *J. Fluorine Chem.*, 2001, **109**, 49.

Radiolabelled amino acids for tumor imaging with PET, J. McConathy, L. Martarello, E.J. Malveaux, V.M. Camp, N.E. Simpson, C.P. Simpson, G.D. Bowers, J.J. Olson and M.M. Goodman, *J. Med. Chem.*, 2002, **45**, 2240.

Fluorinase mediated C–[18]F bond formation: an enzymatic tool for PET labelling, H. Deng, S.L. Cobb, A.D. Gee, A. Lockhart, L. Martarello, R.P. McGlinchey, D. O'Hagan and M. Onega, *Chem. Commun.*, 2006, 652.

[13]N-Labelled L-amino acids for *in vivo* assessment of local myocardial metabolism, F.J. Baumgartner, J.R. Barrio, E. Henze, H.R. Schelbert, N.S. MacDonald, M.E. Phelps and D.E. Kuhl, *J. Med. Chem.*, 1981, **24**, 764.

Synthesis and characterization of iodine-123 labelled 2β-carbomethoxy-3β-(4′-((Z)-2-iodoethenyl)phenyl)nortropane. A ligand for *in vivo* imaging of serotonin transporters by single-photon-emission tomography, M.M. Goodman, P. Chen, C. Plisson, L. Martarello, J. Galt, J.R. Votaw, C.D. Kilts, G. Malveaux, V.M. Camp, B. Shi, T.D. Ely, L. Howell, J. McConathy and C.B. Nemeroff, *J. Med. Chem.*, 2003, **46**, 925.

The biomedical chemistry of technetium and rhenium, J.R. Dilworth and S.J. Parrott, *Chem. Soc. Rev.*, 1998, **27**, 43.

The role of co-ordination chemistry in the development of target specific radiopharmaceuticals, S. Liu, *Chem. Soc. Rev.*, 2004, **33**, 445.

PET imaging of biomolecules using metal–DOTA complexes, K. Tanaka and K. Fukase, *Org. Biomol. Chem.*, 2008, **6**, 815.

[99m]Technetium carbohydrate conjugates as potential agents in molecular imaging, M.L. Bowen and C. Orvig, *Chem. Commun.*, 2008, 5077.

Radiosynthesis and evaluation of [[18]F] Selectflour bis(triflate), H. Teare, E.G. Robins, A. Kirjavainen, S. Forsback, G. Sandford, O. Solin, S.K. Luthra and V. Gouverneur, *Angew. Chem. Int. Ed.*, 2010, **49**, 6821.

# Glossary

**Actinides** Elements from actinium onwards in which the 5f shell is being filled. They are analogous to the lanthanides (rare earths) and are radioactive.

**Activity** In terms of radioactivity, this refers to the number of nuclear transformations in a unit of time. The SI unit is the becquerel (Bq), which is equal to one nuclear transformation per second. The older unit is the curie, which is $3.7 \times 10^{10}$ nuclear transformations per second. The specific activity is a concentration term describing the radioactivity per unit mass, *e.g.* kilobecquerels per milligram or millicuries per milligram.

**Alkaloids** A large family of basic nitrogenous natural products.

**Alpha particle** Radioactive emission consisting of a helium nucleus (two protons and two neutrons) commonly found in the radioactive decay of the heavier elements.

**Auger spectroscopy** The photoinduced ejection of an electron from an atom. This can lead to electron reorganization and the emission of a second electron of characteristic energy.

**Autoradiography** The identification of sites of radioactive decay in an object, *e.g.* a chromatogram, by their effect on a photographic plate.

**Becquerel** The unit of radioactivity equivalent to one nuclear transformation per second. The symbol is Bq. The more common multipliers are kBq or MBq.

The Organic Chemistry of Isotopic Labelling
By James R. Hanson
© James R. Hanson 2011
Published by the Royal Society of Chemistry, www.rsc.org

**Beta particle (β-emission)** An electron ($β^-$) which is emitted from an isotope during a radioactive transformation. The β-particles have a range of energies that are characteristic of the particular isotope involved. Whereas the electron is negatively charged, the positron ($β^+$) is positively charged and of approximately the same mass. The annihilation of a positron by collision with an electron leads to the emission of two photons at 180° to each other and is the basis of positron emission tomography (PET).

**Bremsstrahlung** A form of electromagnetic radiation which arises from the impact of ionizing radiation on material.

**Cahn–Ingold–Prelog sequence rules** Rules for describing the absolute stereochemistry of a chiral centre (see Chapter 4).

**Carbon dating** The long-lived radioactive isotope of carbon ($^{14}C$) is formed in the atmosphere by the reaction

$$^{14}N + n \rightarrow {}^{14}C + {}^{1}H$$

This carbon-14 exchanges with carbon-12 and then it is assimilated by living organisms. However, on the death of the organism metabolism ceases and carbon is no longer assimilated. The carbon-14 in the existing metabolites then decays. Hence the age of the material which had at one time been living can then be established by measuring the residual radioactivity.

**Carrier-free** A preparation of a radioisotope to which no non-radioactive carrier has been added to facilitate isolation and purification. This material will have a relatively high specific activity.

**Chemo-enzymatic synthesis** A laboratory synthesis which involves a combination of chemical and enzymatic steps. The enzymatic step is often concerned with the construction of a chiral centre.

**Chiral** A molecule is chiral if it cannot be superimposed on its mirror image. Although the term strictly applies to the molecule as a whole and includes situations when the molecular dissymmetry arises from restricted rotation, the term is often used loosely to refer to a particular centre.

**Chiral auxiliary** A group which is added to a molecule to make a derivative which is chiral in order to direct the attack of an achiral reagent to one diastereoface of a molecule and thus create a new chiral centre. Once this new chiral centre has been created, the chiral auxiliary may be removed.

**Citric acid cycle** A cyclic sequence of metabolic reactions in primary cellular biochemistry which involves citric acid and leads to the breakdown of acetylcoenzyme A and the release of energy. It is sometimes known as the tricarboxylic acid or Krebs cycle.

**Curie** An earlier unit of radioactivity which is defined as being equivalent to the number of nuclear disintegrations per second (dps) produced by one gram of radium. One curie (1 Ci) is $3.7 \times 10^{10}$ dps. The usual quantities are µCi or mCi.

**Cyclotron** An instrument for accelerating a beam of charged atomic particles in a powerful magnetic field into a spiral by the application of an alternating high-frequency potential difference. The resultant high-energy particles are used to bombard target elements to bring about nuclear reactions.

**Daughter nuclide** A daughter of a nuclide is one that originates from it as a consequence of radioactive decay. This can refer to the isotopes in a radioactive decay series.

**Electron capture** A radioactive transformation in which the nucleus absorbs an electron from an inner orbital. As a consequence, the remaining electrons rearrange to fill the empty shell with the emission of energy as electromagnetic radiation (X-rays) and/or Auger electrons.

**(E) and (Z)** The descriptors for geometric isomers (see Cahn–Ingold–Prelog sequence rules, Chapter 4).

**Exchange reactions** A process in which one isotope in a molecule is replaced by another without the overall disruption of the molecule.

**Gamma (γ) emission** Electromagnetic radiation produced by radioactive decay.

**Geiger–Müller Tube, Geiger counter** A method for detecting and measuring radioactivity by the ionization of a gas that the radiation produces and the consequent discharge of an electric potential between two plates.

**Gray** A unit of the adsorbed dose of radioactivity measured in terns of joules per kilogram.

**Half-life** The time in which the activity of a radioactive isotope decays to half its initial value.

**Heavy water** An older term for deuterium oxide.

**IAEA** International Atomic Energy Agency

**ICRP** International Commission for Radiation Protection

**Isoprenoid natural products** Compounds, typically terpenoids and steroids, whose structures are assembled from $C_5$ isoprene units.

**Isotopes** Nuclides having the same atomic number but different mass number arising from a different number of neutrons within the nucleus. The symbol for the isotope, *e.g.* $^{14}C$, is placed in square brackets immediately preceding the part of the molecule to which it refers. The site of the label, *e.g.* [1-$^{14}C$]acetic acid, is normally included in the bracket.

**Isotope effect** The difference in physical or chemical properties of a compound arising from the substitution of an isotope of a constituent atom by another isotope. Typical properties which show these differences may be boiling point or rates of reactions (see Chapter 3).

**Isotopic abundance** The proportion, usually expressed as a percentage, of the atoms of an element that comprise a particular isotope. Isotopic abundances may vary slightly depending on the source of the element (*e.g.* lead).

**Natural abundance** A term used to describe the natural concentration of a particular isotope. Thus carbon-13 has a natural abundance of 1.1%.

**Neutron** One of the fundamental atomic particles. It has almost the same mass as the proton but is electrically neutral. Consequently, neutron bombardment can penetrate the atom more effectively.

**Nuclide** An individual atomic species characterized by its specific mass number, atomic number and nuclear energy state. The term tends to be used in the context of radioactive decay series where the same isotope may have different nuclear energy states.

**Phosphor** A substance which absorbs radiation and then re-emits the radiation as light and continues to do so from a metastable state for some time after the initiation event has ceased. If the emission ceases immediately after the initiating radiation has ceased, the phenomenon is known as fluorescence.

**Polyketide** A natural product that is formed by the linear assembly of acetate, or sometimes propionate, units.

**Positron** A positively charged β-particle.

**Positron emission tomography (PET)** A three dimensional imaging technique based on locating the origin of photons emitted as a consequence of the annihilation of a positron.

**Rad** An older unit which is used to describe the absorbed dose of radiation (*cf.* gray, Gy); 1 rad is equivalent to 0.01 Gy.

**Radioactive decay** The process by which a radioactive nucleus breaks down with the emission of α- or β-particles or γ-rays. A radioactive decay series links the nuclides that are formed as the radioactive decay of a nucleus progresses.

**Radiochemical purity** The proportion of the radioactivity which is present in a substance in the stated form. A substance may be of high chemical purity but may have a much lower radiochemical purity if the contaminant has a high specific activity. It is important to crystallize a compound to constant specific activity.

**Radiopharmaceutical** A radioactive drug used for diagnostic or therapeutic purposes.

**Rem** This is a unit of dose (Röntgen equivalent man) and is a measure of the exposure (röntgen) multiplied by an absorbing factor which is related to the type of radiation. The sievert (Sv) is the SI unit which has replaced the Rem.

**Scintillators** Compounds which emit a flash of light on impact of ionizing radiation. They play a very important role in the quantification of radiation.

**Single photon emission computed tomography (SPECT)** A three-dimensional imaging technique based on locating the source of photons from an isotope-emitting $\gamma$-radiation.

**Specific activity** The activity per unit mass of a compound. It is usually expressed as $\text{kBq mg}^{-1}$ or $\text{mCi mg}^{-1}$ or per mmol.

**Time-course experiment** A biosynthetic experiment in which the specific activity of a metabolite is plotted against time in an effort to identify biosynthetic sequences.

**Tomography** The measurement of radioactivity within layers in the body, the combination of which is used to construct a three-dimensional image.

# Appendix

## 1 SOME SUPPLIERS OF ISOTOPICALLY LABELLED COMPOUNDS

Although many chemical suppliers list a few compounds labelled with stable isotopes (*e.g.* NMR solvents), the following have a significant list. However, mention of a company in this list does not signify endorsement by the RSC or by the author.

Isotec (Sigma-Aldrich)
3050 Spruce Street, St. Louis, MO 63178, USA

Sigma-Aldrich
New Road, Gillingham, Dorset SP8 4JL, UK

PerkinElmer
940 Winter Street, Waltham, MA 02451, USA
Saxon Way, Bar Hill, Cambridge CB23 8SL, UK
Llantrisant Business Park, Llantrisant CF72 8YW. UK
N.B. GE Healthcare closed its catalogue sales in 2009 and transferred them to PerkinElmer. NEN Radiochemicals are also supplied by PerkinElmer.

Cambridge Isotope Laboratories
50 Frontage Road, Andover, MA 01810, USA
UK distributors are CK Gas Products Ltd, 3 Murrel Green Business Park, Hook, Hampshire RG27 9GR, UK

Icon Isotopes
19 Ox Bow Lane, Summitt, NJ 07901, USA

QMX Laboratories
4 Bolford Street, Thaxted, Dunmow, Essex CM6 2PY, UK

American Radiolabeled Chemicals
101 ARC Drive, St. Louis, MO 63146, USA

A R C UK
28 Aegean Apartments, 19 Western Gateway, London E16 1AR, UK

Medical Isotopes
100 Bridge Street, Pelham, NH 03076, USA
Formerly Stohler Isotope Chemicals

Omicron Biochemicals
115 South Hill Street, South Bend, IN 46617, USA

Paradigm Organics
101 Woodwinds Industrial Court, Suite H, Cary, NC 27511, USA

Vitrax Radiochemicals
660 South Jefferson Street, Unit E Placentia, CA 92870, USA

Moravek Biochemicals
577 Mercury Lane, Brea, CA 92821, USA

Isotopes produced in Russia may be obtained from Isoflex or Ritverc
10 Kurchatov Street, 194223 St. Petersburg, Russia

In addition, there are a number of suppliers of specialist labelled compounds such as amino acids, carbohydrates and environmental standards, such as Wellington Laboratories, CortecNet, AptoChem and ACP Chemicals. A number of the above companies and others, such as Quotient Amersham Bioresearch (Amersham Radiolabeling Services, Amersham Place, Little Chalfont, Bucks HP7 9NA, UK) and BlyChem (Suite 5, Belasis Court, Belasis Hall Technology Park, Billingham TS23 4AZ, UK), undertake custom syntheses. Before ordering and transporting radioisotopes, it is important to check the relevant national regulations covering import/export, transport, storage and disposal of radioactive material. It is also worth bearing in mind that under optimal storage conditions carbon-14-labelled compounds can undergo radiation-induced decomposition to the extent of 1–3% per annum, tritium-labelled compounds can undergo 1–3% decomposition per month and phosphorus-32-labelled compounds 1–2% per week. Almost contrary to chemical intuition, crystalline radio-chemicals may undergo a more rapid radiation-induced degradation than those dispersed in the frozen matrix of a solvent such as benzene. This self-decomposition can be a particular problem with tritium-labelled com-pounds and arises from the formation of hydroxyl radicals.

## 2  LIQUID SCINTILLATION COUNTING

### Some Primary Scintillators

| Compound | Abbreviation | Typical concentration/ g $l^{-1}$ [a] | Fluorescence wavelength/ nm |
|---|---|---|---|
| 2,5-Diphenyloxazole | PPO | 5 | 370.3 |
| 2,5-Diphenyl-1,3,4-oxadiazole | PPD | 7 | 346.6 |
| 2-Phenyl-5-(4-biphenylyl)-1,3,4-oxadiazole | PBD | 12 | 366.9 |
| 2-(4-*tert*-Butylphenyl)-5-(4-biphenylyl)-1,3,4-oxadiazole | Butyl-PBD | 12 | 366.0 |
| *p*-Terphenyl | TP | 7 | 341.8 |

[a]Concentrations in toluene.

### Some Secondary Scintillators

| Compound | Abbreviation | Typical concentration/ g $l^{-1}$ [a] | Fluorescence wavelength/ nm |
|---|---|---|---|
| 2,5-Di(4-biphenyl)oxazole | BBO | 0.5 | 417 |
| 1,4-Bis(2,5-phenyloxazolyl) benzene | POPOP | 0.5 | 423.7 |

[a]Concentrations in toluene.

### Typical Solvents

Toluene, xylene, pseudocumene, dioxane, 2-methoxyethanol.

### Typical Scintillation Cocktails

(a) PPO (5 g), POPOP (0.5 g), AR toluene (1 l).
(b) PPO (7 g), POPOP (0.5 g), AR xylene (650 ml), 2-methoxyethanol (350 ml). The methoxyethanol may be replaced by ethoxyethanol.

(c) For trapping [$^{14}$C]carbon dioxide: PPO (7 g), POPOP (0.5 g) AR xylene (600 ml), 2-methoxyethanol (225 ml), ethanolamine (175 ml). The ethanolamine can be replaced by 3-methoxypropylamine.

(d) *p*-Terphenyl (4 g), POPOP (0.1 g), AR toluene (1 l).

(e) Surface-active agents such as Triton X-100 can be used to enable aqueous solutions to be counted. In all cases calibration curves need to be constructed to allow for quenching.

## 3   PROPERTIES OF ISOTOPES

### Some Magnetic Isotopes in Organic Chemistry

| Isotope[a] | Spin | Natural abundance (%) | Sensitivity (relative to $^1H$) | Frequency at 11.74 T (MHz) |
|---|---|---|---|---|
| $^1$H | 1/2 | 99.98 | 1 | 500.00 |
| $^2$H | 1 | $1.5 \times 10^{-2}$ | $9.65 \times 10^{-3}$ | 76.753 |
| $^3$H | 1/2 | 0 | 1.21 | 533.317 |
| $^7$Li | 3/2 | 92.58 | 0.29 | 194.317 |
| $^{11}$B | 3/2 | 80.42 | 0.17 | 160.419 |
| $^{13}$C | 1/2 | 1.108 | $1.59 \times 10^{-2}$ | 125.721 |
| $^{14}$N | 1 | 99.63 | $1.01 \times 10^{-3}$ | 36.118 |
| $^{15}$N | 1/2 | 0.37 | $1.04 \times 10^{-3}$ | 50.664 |
| $^{17}$O | 5/2 | $3.7 \times 10^{-2}$ | $2.91 \times 10^{-2}$ | 67.784 |
| $^{19}$F | 1/2 | 100 | 0.83 | 470.385 |
| $^{29}$Si | 1/2 | 4.7 | $7.84 \times 10^{-3}$ | 99.325 |
| $^{31}$P | 1/2 | 100 | $6.63 \times 10^{-2}$ | 202.404 |

[a]For a more complete list, see NMR Nomenclature, Nuclear Spin Properties and Conventions for Chemical Shifts (IUPAC Recommendations 2001), R.K. Harris, E.D. Becker, S.M. Cabral de Menzes, R. Goodfellow and P. Granger, *Pure Appl. Chem.*, 2001, **73**, 1795–1818.

### Stable Isotopes Commonly Detected by Mass Spectrometry

| Isotope | Atomic weight ($^{12}C = 12.000$) | Natural abundance (%) |
|---|---|---|
| $^1$H | 1.007825 | 99.985 |
| $^2$H | 2.014102 | 0.015 |
| $^{13}$C | 13.003354 | 1.108 |
| $^{14}$N | 14.003074 | 99.64 |
| $^{15}$N | 15.000108 | 0.37 |

(*Continued*).

| Isotope | Atomic weight $(^{12}C = 12.000)$ | Natural abundance (%) |
|---|---|---|
| $^{16}$O | 15.994915 | 99.8 |
| $^{17}$O | 16.999133 | 0.037 |
| $^{18}$O | 17.999160 | 0.2 |
| $^{19}$F | 18.998405 | 100 |
| $^{23}$Na | 22.9897 | 100 |
| $^{29}$Si | 28.976927 | 92.2 |
| $^{30}$Si | 29.973761 | 4.7 |
| $^{32}$S | 31.972074 | 95.0 |
| $^{34}$S | 33.967865 | 4.2 |
| $^{35}$Cl | 34.968896 | 75.53 |
| $^{37}$Cl | 36.965896 | 24.47 |
| $^{79}$Br | 78.918348 | 50.54 |
| $^{81}$Br | 80.916344 | 49.46 |
| $^{127}$I | 126.904352 | 100 |

# Subject Index